W9-BLN-658

Linear Panel Analysis

Models of Quantitative Change

This is a volume of

Quantitative Studies in Social Relations

Consulting Editor: Peter H. Rossi, University of Massachusetts, Amherst, Massachusetts

A complete list of titles in this series appears at the end of this volume.

Linear Panel Analysis

Models of Quantitative Change

Ronald C. Kessler

Department of Sociology
University of Michigan
Ann Arbor, Michigan

David F. Greenberg

Department of Sociology
New York University
New York, New York

ACADEMIC PRESS

A Subsidiary of Harcourt Brace Jovanovich, Publishers

New York   London   Toronto   Sydney   San Francisco

ACADEMIC PRESS, INC.
111 Fifth Avenue, New York, New York 10003

United Kingdom Edition published by
ACADEMIC PRESS, INC. (LONDON) LTD.
24/28 Oval Road, London NW1 7DX

Library of Congress Cataloging in Publication Data

Kessler, Ronald C.
    Linear panel analysis.

    (Quantitative studies in social relations)
    Bibliography: p.
    1. Panel analysis. 2. Social sciences--Mathematical
models. 3. Social change--Mathematical models.
I. Greenberg, David F. II. Title. III. Series.
H61.K47   300'.724   81-3504
ISBN 0-12-405750-0   AACR2

PRINTED IN THE UNITED STATES OF AMERICA

81 82 83 84   9 8 7 6 5 4 3 2 1

Contents

Preface

This book is concerned with important current issues having to do with the use of linear models in the analysis of change data measured on a sample of individuals (or other units of analysis) over multiple time points. Although a great many books have reviewed general issues in the analysis of linear models, this is the first systematic treatment of special issues that arise in linear models of change.

The treatment in the following chapters assumes that the reader has some familiarity with linear regression models as they are used in the social sciences. With that much background it should be possible to follow all of the major arguments in the book. And mastery of these arguments will provide all of the tools needed to carry out an analysis of change data.

This book was written for use by social scientists who work with change data. But we hope it will also prove to be useful for readers who wish to understand the complexities of over-time data analysis.

After a general introduction to the topic of change analysis in Chapter 1, we turn in Chapter 2 to a presentation of some basic algebraic results on change scores. These results will be needed for the remainder of the book. In Chapter 3 we construct and analyze structural equation models for studying the causes of change, while Chapter 4 presents several ways of simply describing change in groups. Chapter 5 discusses the uses and abuses of cross-lagged panel correlations, a widely used but poorly understood technique for analyzing the causal relationships between two variables in panel data.

Chapters 6 through 10 deal with a variety of special topics that arise in panel analysis. In Chapter 6 we introduce methods of solving the identification problem that arises when linear models are used to estimate the effects of initial position from the effects of change in regression analysis. Chapter 7 is concerned with the question of handling serially correlated error terms. Whereas the earlier chapters analyzed *given* panel models, Chapter 8 deals with the question setting up an initial model. In Chapter 9 we examine the sensitivity of parameter estimates to certain forms of model misspecification. Methods for ascertaining the presence of measurement error and making adjustments for it are treated in Chapter 10.

Chapter 11 takes up two alternatives to the analysis of over-time data for a set of individuals: the pooling of cross-sections and time series, and continuous-time models. Finally, in Chapter 12, we discuss some of the practical aspects of designing and carrying out the data collection phase of a panel study.

We are grateful to our research assistants, Roger Brown, Sandy Engel, and Sheri Prupis, for carrying out computations, and to Duane Alwin, Alfred Blumstein, Jacquelyn Cohen, Sara Beck Fein, and Ronald Milavsky for helpful comments and suggestions. Part of this research was funded under Grant 79-NI-AX-0054 from the National Institute of Law Enforcement and Criminal Justice. Points of view are those of the authors and do not necessarily reflect the position of the U.S. Department of Justice. Support was also received under Grants 1 R03 MH32490-01 and 1 R03 MH33927-01 from the National Institute of Mental Health.

1

Introduction

The analysis of change is central to the social sciences. Whether we are concerned with attitudes, beliefs or behaviors, whether with individuals, institutions or whole societies, the central concerns of social scientists revolve around the description of patterns of stability and change, and the explanation of how and why change takes place or fails to take place.

Given this pervasive concern with change, we might expect that methods for analyzing change would have been fully worked out long ago. Yet, while methodologists have developed a variety of approaches for analyzing change, a great deal of confusion continues to surround this subject. Here, our goal in writing about change is to clear away some of this confusion.

Several major methodological statements have argued that nothing "special" need be done in analyzing data regarding change. The rationale for this position is that conventional methods of analysis can be applied because all causally modeled data can be interpreted *as if* actual change were being studied, even when the data have been gathered at a single point in time (Cronbach and Furby, 1969; Goldberger, 1971). In part this is true. The methods we describe here bear many points of resemblance to those appropriate for analyzing data collected at a single point in time. However, certain problems arise when data are collected over time that are not normally encountered when dealing with static data. And, in addition, certain opportunities arise. These problems and opportunities are the subject of this work.

Methods for Analyzing Change

Inferences about change, its determinants and its consequences, are conventionally obtained by means of five general analytical strategies: cross-sectional analyses, trend studies, time series analyses, panel designs, and continuous-time approaches.

Cross-Sectional Analysis

In cross-sectional analyses, change effects are inferred from variations between units at a single point in time. While valuable insights can often be gained from such analyses, this approach can be seriously flawed. For example, let us suppose that in a cross-sectional survey, young people are found to endorse more liberal positions on political issues than older people One interpretation is that as people age they become more conservative. But another interpretation is also possible. If one's position on a conservatism–liberalism spectrum is determined for life by political events that occur in one's youth, and the older people in the sample were exposed to different events than the younger people, we might see the same pattern. The first of these two interpretations could be distinguished from the second only by following a set of individuals for some period of time to see whether their views changed as they got older.

For this reason, cross-sectional analyses should be avoided if possible whenever cohort or historical effects are thought to exist, at least when concern centers on change.[1]

Trend Studies

When we are content with simply determining by how much a single variable has changed in a population, we can carry out a trend study. For example, pollsters may want to know whether the proportion of the electorate planning to vote for a candidate has increased or decreased as the date of election draws near, without caring about the reasons for a change on the part of any specific individual. Economists may want to know how average net income has changed in the population over time, while remaining indifferent to the question of whether the income of some specific

---

[1] A distinction can be made between a cross-sectional strategy for *collecting* data and a cross-sectional *analysis*. In survey research, for example, one may ask questions of a sample at a single point in time, but these questions may elicit information about individuals' opinions, attitudes, or behavior at earlier times. Taking these responses as (possibly imperfect) indicators of prior levels of variables makes it possible to analyze genuine change.

individual has changed relative to the income of others. In problems such as these, the desired information can be obtained efficiently by drawing successive samples from the same population at different points in time.

Time Series and Panel Designs

Social scientists are often concerned with differences between units in patterns of change, or in the effects of a change in one variable on another variable. We might ask, for instance, if people change their political party affiliations when their incomes increase, or whether worker efficiency is improved when management plays Muzak over a loudspeaker system. The investigation of such questions is aided by the collection of data at more than one point in time on the *same* units of analysis. The most widely utilized design of this type is the time series, where observations are collected on a single individual or other unit of analysis for a large number of time intervals (Hibbs, 1974; Box and Jenkins, 1976). The sequences of annual economic indicators for the economy (GNP, unemployment rate, etc.) are probably the most familiar time series to most of us; here the unit of analysis is the national economy.

A second method is the panel study, where observations are collected for a number of individuals (or other units of analysis) instead of only one, at two or more points in time.[2] This design is especially powerful in that it provides information about cross-sectional as well as longitudinal variation. In addition, it allows us to study variations between individuals in patterns of change.

Continuous-Time Designs

All four of the above designs (cross-sectional, trend, time series, panel analysis) are what Campbell (1978) has called "snapshot" designs; they record the state of our units of analysis at one or more discrete moments, but without providing information about the patterns of movement that take place between successive "frames" of the static "pictures." Often this is the only way to study change.

Yet at times information is available about the state of each unit at every moment, or about the precise timing of events. For example, we may have a record of someone's income, occupational or marital status, not just at discrete intervals, but at all times. The exact moment at which

[2] The collection of a time series of observations on a number of units of analysis is often called multiple time series, or a multiple-wave panel design. This approach is simply a panel study with data collected at an unusually high number of time points.

a riot breaks out may be a matter of record.[3] Here we obtain information about the ongoing pattern of stability and change in the variable of interest. We have a "motion picture" instead of a succession of snapshots (as in film, the distinction becomes blurred when the snapshots are taken frequently enough; after all, a film strip is really a sequence of "stills").

When information of this kind is collected, time is treated as a continuous rather than a discrete variable. Campbell characterizes this approach as a "life history" design, but the approach is by no means restricted to the study of individual life histories. When suitable records can be located in archives, continuous-time designs can be used to analyze change in organizations or communities. For example, Tilly *et al.* (1975) have studied patterns of collective violence by analyzing the dates and durations of various forms of civil strife in European political units over a period of several decades.

This book is concerned with models for the analysis of quantitative change, with particular attention to methods for the causal analysis of change. Since methods for analyzing time series are already familiar with social scientists, we will be concerned here primarily with panel designs. Continuous time designs will be discussed as well, but much more superficially. Our treatment of continuous time designs is brief because social scientists only infrequently have suitable data, and because the mathematical background required for this type of analysis greatly exceeds that of most social scientists.[4]

The Historical Development of Panel Analysis Methods

Panel analysis was introduced into the social sciences in several contexts. Psychometricians have long carried out panel analyses in studying the reliability of measurement. Since personality traits are expected to be highly stable over time, an obvious procedure for evaluating instruments for measuring personality is to verify that they yield scores for individuals that are highly stable over time. Neither a trend study nor a time series analysis would provide this information; thus it was natural for psychometricians to turn to panel analyses for this purpose.

---

[3] Where the levels of variables of interest are not routinely recorded on an ongoing basis, this analysis will generally not be possible. Except where diaries are kept, for example, we would have no continuous record of people's moods. At best we would be able to do a panel study by asking them about their moods from time to time.

[4] A more detailed discussion of various forms of continuous time designs can be found in Hannan and Tuma (1979, forthcoming).

A second context was the series of child development studies begun under the auspices of the National Research Council in the mid 1920s. A number of these studies were designed to be, and succeeded in being, "womb-to-tomb" panels (Sontag, 1971), indexing patterns of development among children at various ages.[5]

A third context involved the modeling of changes of brand preference in marketing studies and, somewhat later, in studies of political preference. These studies, all inspired by the work of Paul Lazarsfeld (Lazarsfeld *et al.*, 1948; Glock, 1955), had two sorts of concerns. One had to do with changes in a single variable. For example, does party affiliation change over time, and if so, how? What is the nature of the process that generates occupational mobility? In analyzing processes of this kind, the variable of interest was typically dichotomized and efforts were made to find a mathematically simple stochastic (probabilistic) model that would fit the cell frequencies representing transitions from one time to another reasonably well. Early models of this kind are discussed by Wiggins (1973) and Coleman (1964); more recent discussions will be found in Singer and Spilerman (1976, 1979).

The second concern of the marketing tradition involved the experimental or quasi-experimental analysis of the relationships between categorical (usually dichotomized) variables. The early market studies were as concerned as those of today in relating changing preferences to the characteristics of individual consumers. During World War II, researchers studied the cognitive impact of propaganda films. Out of this concern came Lazarsfeld's famous method for separating contending causal hypotheses through the analysis of the 16-fold table (Lazarsfeld, 1948; Kessler, 1977a).

Recent work on log-linear models for contingency table analysis (Goodman, 1975; Duncan, 1980) has revived interest in this general approach, and has facilitated extension to multivariate, nondichotomized contingency tables. Yet corrections for unreliability owing to measurement error are difficult in this approach. The introduction of control variables can lead to a rapid reduction in cell frequencies, posing problems for statistical analysis. And where one's variables are truly interval-level, the contingency-table approach requires the introduction of arbitrary cutting points, and the loss of information contained in the data. These limitations have been a major stimulus to the development and use of alternative analytical methods.

---

[5] These studies and their implications for the understanding of behavior problems in children are reviewed in Robins (1979a).

Linear Models for Analyzing Panel Data

While panel studies date back to the beginnings of modern-day survey research, they are currently the subject of intense interest in the social science research community. One of the reasons for this surely involves the growth of interest in social change, and the corollary loss of interest in equilibrium models of society. Intersecting with this theoretical shift has been the introduction of structural equation methods for analyzing causal relations among interval-level variables. Following their introduction in the late 1960s (Duncan, 1966a), these methods quickly gained wide acceptance as analytical tools.

As social scientists came to understand structural equation methods, they began to realize, as economists had some time earlier, that these methods (like other statistical techniques) could be used to reach conclusions about dynamic processes on the basis of cross-sectional data only under rather restrictive circumstances.

This realization led the more naive users of structural equations to turn to panel studies in the hope that over-time data could unequivocally establish the causal priority of variables, and thus make it possible for researchers to carry out causal analyses free from any a priori assumptions. More sophisticated methodologists quickly demonstrated that this is not the case. Yet the uninformed continue to plunge ahead on the assumption that panel models will wash away all methodological difficulties, and find it a shock when they discover that this is far from true.

Even as the limitations of linear models for panel data have become apparent, methodologists have found them to be extremely useful in specifying the sorts of dynamic processes that are implied by many sociological theories. A correct specification of causal relations frequently requires lagged data. Similarly, the separation of age, period, and cohort effects is accomplished more easily with through-time data, though certain assumptions about the causal processes at work must still be made.

2

The Algebra of Change

To study change quantitatively in a sample of cases, we must be able to define change precisely enough so that we can say how much a variable has changed over a period of time, and what the causes of that change are. For a single case i, the quantity $\Delta X_i = X_{it} - X_{it'}$ is a natural measure of the extent to which X_i has changed between time t' and time t. The set of values ΔX_i for the entire sample provides complete information about change in X, but for many purposes would be overly informative. To answer a question such as "Overall, has X changed a little or a lot?" we want a single statistic that measures change in the sample, not an entire set of statistics for each case. Our first task will be to develop a suitable measure of change from which we can derive results that will aid us in understanding the causes of change. These results we obtain for linear models of change are extremely useful in the analysis of panel data, as we will demonstrate by using them to interpret two previously published panel studies. The equations derived in this chapter will also be needed for the developments presented in later chapters.

Stability and Change

Conceptually, stability is the opposite of change. To say that a variable is stable over time is to say that it does not change very much. For this reason it should be possible to translate mathematical statements about the stability of a variable into statements about change without much difficulty. This interchangability between statements concerning change and those concerning stability proves to be a great convenience in carrying out many analyses.

To see how we can go back and forth between stability and change, we assume that a single variable X is measured at times 1 and 2 for a number of different cases i. More than a decade ago, Heise (1969) proposed that the stability of X could be measured conveniently by the Pearson correlation coefficient $r_{X_1X_2}$. The appeal of the correlation coefficient as an indicator of stability rests on the intuitive thinking that if $r_{X_1X_2}$ is close to 1 after correction for measurement error, then X has not changed much during the interval. This approach implicitly treats the correlation coefficient as an inverse measure of change. To translate the inverse measure into a direct measure of change we can use any function of $r_{X_1X_2}$ that decreases steadily as $r_{X_1X_2}$ increases. The quantity $-r_{X_1X_2}$ would be an especially attractive choice since it is linear in $r_{X_1X_2}$ and has the same range (-1 to $+1$).

A moment's thought shows that the intuitive reasoning suggesting that the correlation coefficient measures stability is fallacious, for it is not necessarily true that X has changed by a little whenever $r_{X_1X_2}$ is close to 1. As long as X_2 is accurately predicted from X_1, the correlation between these two variables will be close to 1 regardless of how much or how little X has changed.

Recently Wheaton et al. (1977) have argued that in the multivariate case the partial regression coefficient should be used as a measure of stability, rather than the correlation coefficient. The partial regression coefficient, they suggest, should be interpreted as the extent to which X_1 is a "source" of X_2:

> We contend that the simple change definition (operationalized as a correlation) can lead to misleading interpretations of stability. . . . Stability, as we define it, is concerned with the amount of change or lack of change in X_{t+1} and due to X_t alone; that is, the degree to which one's score on X_t is, in fact, the source of one's score on X_{t+1}. . . . Such a notion of stability is implicitly recognized by Heise when describing potential sources of distortion to stability estimates [Wheaton et al., 1977, p. 91].

The notion that the value of a variable at one time is a source of the value of that variable at a later time is recognized in the popular adages that "success breeds success" and "the rich get richer and the poor get poorer," as well as in the Biblical warning that "Unto every one that hath shall be given, and he shall have abundance; but from him that hath not shall be taken away even that which he hath [Matthew 25:29]."

Once we think of time 1 variables as a source for time 2 variables, refinement in the analysis of stability and change becomes possible, for there are actually two quite different ways in which X_t can be a "source" of X_{t+1}. The first is that X has not changed much between time t and time $t + 1$, so that X_t and X_{t+1} are almost identical. The second is that X_{t+1}, though quite different from X_t, is largely *determined by* X_t. For instance, it is one thing to say that the prestige of a person's job at age 25 remains *totally unchanged* throughout that person's subsequent occupational career, and quite another to say that the prestige of the job at age 25 *totally determines* subsequent prestige changes. The first situation is one of stability, while the second involves highly structured change.

Insight into the contribution that stability and structured change make to the overall correlation $r_{X_1 X_2}$ can be gleaned from an algebraic investigation. For person i in the population, the time 2 score X_{i2} can be expressed as the sum of the initial score X_{i1} and the change ΔX_i:

$$X_{i2} = X_{i1} + \Delta X_i. \tag{2.1}$$

Summing over individuals and dividing by the number of individuals N, we can express the covariance of initial and later scores in terms of the two component parts of X_2:

$$\text{cov}(X_1, X_2) = s_{X_1 X_2} = s_{X_1}^2 + s_{X_1 \Delta X} = \text{var}(X_1) + \text{cov}(X_1, \Delta X). \tag{2.2}$$

The two terms of this equation express the two ways that X_1 can influence X_2. The first component, $s_{X_1}^2$, is the variance of X at time 1. The extent to which this initial variance makes up the covariance with X_2 is a measure of lack of change in X over time. The second component, $s_{X_1 \Delta X}$, is the covariance of X_1 with the change in X, and measures the causal influence of X_1 on the change part of X_2. Since the correlation coefficient $r_{X_1 X_2}$ is obtained from the covariance $s_{X_1 X_2}$ by dividing the latter by the quantity $s_{X_1} s_{X_2}$, we have succeeded in decomposing the correlation coefficient in a way that shows the relative contributions of stability and structured change to the correlation.

This two-part influence can be seen even more easily in the regression of X_2 on X_1. We begin by writing the regression equation in the familiar form (with index i suppressed):

$$X_2 = a + b_1 X_1 + b_2 Y + e, \tag{2.3}$$

where Y is a variable that influences X, and e is an error term. Subtracting X_1 from both members of the equation gives us an expression for the regression of change in X on X_1. We then have

$$\Delta X = X_2 - X_1 = a + (b_1 - 1)X_1 + b_2 Y + e \tag{2.4a}$$

$$= a + b_1^* X_1 + b_2 Y + e, \tag{2.4b}$$

where b_1^* is the regression of ΔX on X_1 when Y is controlled. Reexpressing

Eq. (2.3) in terms of the parameter b_1^*, we have

$$X_2 = a + (b_1^* + 1)X_1 + b_2Y + e. \qquad (2.5)$$

This equation shows quite clearly the contributions of stability and change in the influence of X_1 on X_2. That the coefficient b_1 represents the influence of X_1 on change in X is evident from Eq. (2.4b). The influence that X_1 has on X_2 as a result of the stability, or lack of change in X, is given by the addition of the constant 1 to the coefficient b_1^*. In the absence of any change in X, this constant term is all we would have. The Appendix to this chapter investigates in some detail the way the correlation coefficient $r_{X_1X_2}$ can be influenced by these two components.

Taken together, these results show that conventional regression equations can be interpreted in terms of change scores. One begins by regressing X_2 on X_1 and some set of other variables Y. To obtain the effect of X_1 on ΔX, subtract 1 from the partial regression coefficient representing the effect of X_1 on X_2 with Y held constant. To obtain the effect of Y on ΔX, take the regression coefficient for the effect of Y on X_2 with X_1 held constant.[1]

Should one wish to go further and examine the effect of a *change* in Y on the change in X, these results can be readily extended. By making use of the identity $Y_2 - Y_1 = \Delta Y$, we can transform the static score equation

$$X_2 = b_{X_2X_1.Y_1Y_2}X_1 + b_{X_2Y_1.X_1Y_2}Y_1 + b_{X_2Y_2.X_1Y_1}Y_2 + e$$

to an equation involving differences:

$$\Delta X = (b_{X_2X_1.Y_1Y_2} - 1)X_1 + (b_{X_2Y_1.X_1Y_2} + b_{X_2Y_2.X_1Y_1})Y_1 \\ + b_{X_2Y_2.X_1Y_1}\Delta Y + e. \qquad (2.6)$$

An equation of this sort implicitly assumes the existence of a cross-sectional causal relationship between X and Y. To say that ΔX is partly determined by ΔY is to say that X_2 is partly determined by Y_2. This kind of

[1] Jackman (1980) has pointed out that in cross-national research, the variables employed (such as population, or per capita gross national product) may be badly skewed. When this is so, error terms in Eq. (2.3) or Eq. (2.4b) may be heteroskedastistic. Under this circumstance, parameter estimates will be inefficient and significance tests will be biased. Jackman points out that a log transform may eliminate much of the skewness in variables. Instead of estimating Eq. (2.4b), one estimates a comparable equation involving the transformed variables:

$$\log(X_2) - \log(X_1) = a + b_1^* \log(X_1) + b_2Y + e.$$

Since $\log(X_2) - \log(X_1) = \log(X_2/X_1)$, the coefficient b_1^* in this approach represents the effect of X_1 on the percentage rate of growth in X_1. Our discussion will assume that error terms are homoskedastistic; however, when a test for heteroskedasticity of error terms indicates that this assumption is invalid, variables should be transformed before proceeding.

cross-sectional influence is not present when only lagged predictors of change are considered.

Another noteworthy feature of Eq. (2.6) is that of the three possible predictor variables Y_1, Y_2, and ΔY, only two may be used in the regression.[2] This is so because, by definition, $\Delta Y = Y_2 - Y_1$. This becomes especially important in substantive applications where we have reason to believe that the three have independent effects. For instance, the literature on mental illness suggests that social class origins (Y_1), social class destination as an adult (Y_2), and experiences of social mobility (ΔY) each independently influence current mental health (Kessler and Cleary, 1980). But so long as one considers only linear models, these three effects cannot be disentangled. In Chapter 6 we show how more complex models can be estimated in a way that separates these three independent influences from one another.

The equivalences between static and change score regressions derived above all apply to *unstandardized* regression coefficients. To standardize these relationships, multiply them by the ratio of the standard deviations of predictors and change scores. For example,

$$\beta_{\Delta X, X_1 \cdot Y} = (b_{X_2 X_1 \cdot Y} - 1)(s_{X_1}/s_{\Delta X}), \tag{2.7a}$$

$$\beta_{\Delta X, Y \cdot X_1} = (b_{X_2 Y \cdot X_1})(s_Y/s_{\Delta X}). \tag{2.7b}$$

This procedure requires that the variance of the change score be calculated. This can be computed directly, or it can be expressed in terms of the static score variances and covariances through the formula

$$s_{\Delta X}^2 = s_{X_1}^2 + s_{X_2}^2 - 2s_{X_1 X_2}. \tag{2.8}$$

These results not only provide a strategy for clarifying the dynamics of change; they also demonstrate that linear difference equations (those involving ΔX as dependent variable) and equations involving only static scores (X_1 and X_2) are mathematically equivalent and can readily be transformed into one another. The commonly asserted claim that partial regression or correlation methods are *mathematically* superior to those involving change scores is entirely vacuous (Bohrnstedt, 1969; Markus, 1980).

However, at times there will be practical reasons for using one or the

---

[2] Which two variables are included is a decision that must be made on theoretical grounds. In the example given, we have chosen to include Y_1 and ΔY, but in particular applications, it may be more reasonable to include Y_2 and ΔY, for example, on the grounds that the interval between time 1 and time 2 is too great for lagged effects to be appreciable. The considerations that lead the researcher to one specification instead of another are discussed further in Chapter 6.

other formulation. In some cases, extremely high zero-order correlations between X_1 and X_2 in a matrix of correlations will make it desirable to eliminate the trivial part of this association by taking first differences. In other cases, one might prefer static scores, as when making use of measurement error adjustment procedures that require certain constraints to be made on the correlations of unobserved to observed scores across time. Models of this sort will be discussed in Chapter 10.

Modeling and Estimating Change

Although the manipulations outlined above are algebraically trivial, they may nevertheless seem a bit confusing to readers who are accustomed to cross-sectional regressions. For example, we have utilized the time 1 variable X_1 to predict change in X. This is not done in cross-sectional analyses because change itself never appears as a variable in cross-sectional analyses.

Substantively, there can be good reasons for including X_1 as a predictor of ΔX (or X_2). In economics, if the rate of interest banks pay on savings deposits is the same for all deposits regardless of how large they are, then, assuming there are no withdrawals, the size of the deposit after some period of time will be proportional to the amount originally deposited. In educational psychology we might find that the arithmetic skills of children who are proficient at arithmetic improve more over the course of a school year than the skills of children who begin the year with poor arithmetic skills. In demography, if isolated societies all have the same birth and death rates, the numerical growth of population in each society will be proportional to its original population.

In all these examples (and many others can easily be imagined), high initial scores on a variable are associated with especially large increases in that variable. But as Coleman (1968) points out, the opposite can also occur. It is not uncommon for a variable whose value is especially high to elicit a response from the environment that tends to reduce the high value. If the environmental factors responsible for this negative feedback are known, they can be included explicitly in the model. Commonly, though, some of these factors will not be known. The inclusion of X_1 in the equation for ΔX is a way of picking up these "control" effects.

Even when a *substantive* reason for including X_1 in the regression equation for X_2 is lacking, there may be a *statistical* reason for doing so. Imagine that the causal dynamics linking X and Y are correctly specified by the following equation:

$$X_t = a_t + bY_t + cZ + e_t. \tag{2.9}$$

Here Z is an unmeasured cause of X. By omitting the subscript t from Z, we imply that it remains constant between times 1 and 2. We can think of Z as a contribution to X that is unique for each individual case, and unchanging in the period under consideration.[3]

If Z were an observed variable, Eq. (2.9) could be estimated from cross-sectional data at either time 1 or time 2, but we have assumed that Z is not measured. On the other hand, if we forget about Z and try to estimate b by regressing Y on X using cross-sectional data, we would run the risk of specification bias. To avoid this risk, we eliminate Z from the regression by subtracting the time 1 equation from the time 2 equation:

$$X_2 - X_1 = (a_2 - a_1) + b(Y_2 - Y_1) + e_2 - e_1. \qquad (2.10)$$

The parameter b can now be estimated without bias from the difference scores, though of course the parameters a_1, a_2, and c cannot be estimated.

It is sometimes convenient to rewrite Eq. (2.10) as

$$X_2 = X_1 + (a_2 - a_1) + b(Y_2 - Y_1) + e_2 - e_1. \qquad (2.11)$$

This is an equation of the same general form as Eq. (2.3), but with the coefficient of X_1 fixed at the numerical value of 1. The computer program LISREL, which is especially convenient for carrying out a number of the procedures we describe in subsequent chapters (and is described in Appendix B to this chapter), contains an option that permits regression coefficients to be fixed at arbitrary values of this sort. Nevertheless, we caution that estimation of the parameter b is not as straightforward when working with Eq. (2.11) as it is when working with Eq. (2.10). The difficulty with Eq. (2.11) is that X_1 is correlated with e_1. This means that if X_1 is used to identify Eq. (2.11), parameter estimates will be biased.[4] This problem does not arise in working with Eq. (2.10) since both e_1 and e_2 are assumed to be uncorrelated with Y_1 and Y_2. So this parametrization is to be preferred for purposes of estimation.

To illustrate the computations we have been discussing and to show how findings are interpreted, we consider a two-wave panel study of the relationship between aggressive behavior and the viewing of violent television shows. Lefkowitz et al. (1977) measured preference for violent television shows (TV) and peer-rated aggressive behavior (AG) for 184 boys in 1960 and again in 1970. The observed correlations and standard deviations they obtained are shown in Table 2.1.

---

[3] This model and the statistical properties of its estimators are discussed in greater detail in Chamberlain (1980).

[4] An unbiased estimate of b could, however, be obtained from Eq. (2.11) by using the quantity $(Y_2 - Y_1)$ as an instrument.

Table 2.1

Correlations and Standard Deviations for Boys'
Preference for Violent Television Programs and
Their Aggressive Behavior[a]

TV1	5.05	.05	.21	.31
TV2		6.09	.01	−.05
AG1			11.25	.38
AG2				9.86

Source: Lefkowitz *et al.* (1977, pp. 73–74, 128–129).

[a] TV = preference for violent television programs; AG = peer-rated aggressive behavior. Entries on diagonals are standard deviations; those on off-diagonals are correlation coefficients. Sample size is 184.

In all likelihood, the causal relationship between a preference for violent television shows and aggressive behavior is complex. To simplify the analysis, we suppose that TV may influence AG, but that AG has no influence on TV. Also, though aggressive behavior is undoubtedly influenced by many variables besides preference for violent television shows, we neglect these other variables to simplify the exposition. Our concern here is not with the substantive question of what causes aggressive behavior, but with illustrating a method for analyzing data. We suppose, then, that aggression at time 2 depends linearly on aggression at time 1 and on preference for violent television shows at time 1. The estimation is straightforward, and leads to the following prediction equation:

$$\hat{AG}_2 = 4.29 + .28A_1 + .47TV_1$$

in unstandardized form. In standardized form this becomes:

$$\hat{ag}_2 = .33ag_1 + .24tv_1.$$

Since the coefficient of TV_1 is positive, we conclude that preference for violent television shows at time 1 tends to increase boys' subsequent aggressive behavior. We note also that the quantity (.28-1) is negative, indicating that boys who are especially aggressive at time 1 tend to decline in aggressiveness more than boys who are initially less aggressive. Nevertheless, since the coefficient of A_1 is positive, we conclude that initially aggressive boys are still more aggressive at time 2 than boys who are initially not aggressive. Even though the aggressiveness of the highly aggressive boys declines by a larger amount, it does not do so by enough to eliminate the positive relationship between aggressiveness at the two times.

If we alter the assumptions of the model just considered to suppose that aggressive behavior is influenced primarily by instantaneous preference for violent television shows, and by an unmeasured, idiosyncratic variable (perhaps biological or psychological in origin) that does not vary over time, then we are led to estimate the equation

$$AG_2 - AG_1 = b(TV_2 - TV_1) + e_2 - e_1.$$

As long as television preferences are not influenced by aggressive behavior, the estimation is straightforward. The least squares estimate of b is

$$\hat{b} = \text{cov}\{(AG_2 - AG_1), (TV_2 - TV_1)\}/\text{var}(TV_2 - TV_1)$$
$$= \frac{\{\text{cov}(AG_2,TV_2) - \text{cov}(AG_1,TV_2) - \text{cov}(AG_2,TV_1) + \text{cov}(AG_1,TV_1)\}}{\text{var}(TV_2) + \text{var}(TV_1) - 2\,\text{cov}(TV_2,TV_1)}$$
$$= -.11.$$

The corresponding standardized coefficient is $\hat{\beta} = -.07$.

We call attention to the fact that in this second model we find a small, negative estimate for the effect of television viewing preference on aggressive behavior. This contrasts with a moderate and positive effect in the model previously considered. The discrepancy between the two estimates underscores the importance of correctly specifying the causal relationships at work. Each of the models we estimated is based on specific assumptions about how preference for violent television shows affects aggressive behavior. The confidence one would have in the parameter estimates for the two models can only be as good as the confidence one would be willing to place in the untested assumptions of the two models.[5]

[5] The model with an omitted, unmeasured variable can actually be estimated under somewhat more general conditions than the model specified in the text. Suppose that Eq. (2.9) holds, and that Z varies with time. We can always write $Z_{2i} = dZ_{1i} + u$. If u is uncorrelated with Z_1, Y_1, Y_2, e_1, and e_2 (the standard assumptions in regression analysis), we can eliminate Z by multiplying the equation for X_1 by d and subtracting it from the equation for X_2. If variables are standardized, the resulting equation is

$$X_2 = dX_1 + bY_2 - bdY_1 + cu + e_1 - de_1.$$

We cannot use X_1 to identify this equation because it is correlated with the error term, but we obtain two independent equations for the parameters b and d by taking covariances with Y_1 and Y_2. The resulting equations are nonlinear in the parameters and must be solved numerically. When this procedure is carried out using the correlations from Table 2.1, we obtain the estimates $d = 1.15$, $b = -.064$. The estimate of d is somewhat disturbing since it should be bounded above by 1, but it is at least somewhat reassuring that the discrepancy is not large. The estimate of b is close to that found for the model discussed in the text. The statistical properties of these estimators (e.g., sampling distributions) are a subject for future investigation.

Change as a Determinant of Time 2 Status

Additional insight into the interpretation of a regression coefficient as a measure of stability and the relationship of stability to change can be obtained by interpreting Eq. (2.1) as a path equation expressing the separate effects of X_1 and ΔX on X_2. A path diagram exhibiting these influences is given in Figure 2.1, with *unstandardized* coefficients linking X_2 with X_1 and with ΔX each set equal to 1.

The standardized path coefficients (β) can be obtained from the unstandardized coefficients (b) in the usual manner:

$$\beta_{X_2X_1.\Delta X} = b_{X_2X_1.\Delta X}(s_{X_1}/s_{X_2}) = s_{X_1}/s_{X_2}, \qquad (2.12a)$$

$$\beta_{X_2\Delta X.X_1} = b_{X_2\Delta X.X_1}(s_{\Delta X}/s_{X_2}) = s_{\Delta X}/s_{X_2}. \qquad (2.12b)$$

This computation tells us how much of the effect of X_1 on X_2 (as measured by the standardized regression coefficient $\beta_{X_2X_1} = r_{X_2X_1}$) is due to the direct (lack of change) effect of X_1 on X_2.

From Eq. (2.12a) we see that the direct component will be greater than or equal to 1 whenever $s_{X_1}^2 > s_{X_2}^2$. It is quite meaningful to find a direct stability effect that exceeds 1. When this happens, factors leading to change in X are causing initial differences in the population to decrease with time,[6] so that there is a negative covariance between X_1 and ΔX. This phenomenon is known technically as "regression to the mean." We saw an example in our analysis of the Lefkowitz data on television viewing and aggression earlier in this chapter.

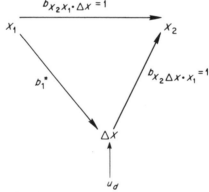

Figure 2.1. The direct and indirect effects of X_1 on X_2.

---

[6] This can easily be shown by expressing the variance of X_2 in terms of the variance of X_1, the variance of ΔX, and the covariance of X_1 and ΔX. On the other hand, when $s_{X_1} < s_{X_2}$, initial differences in the population may be increasing or decreasing, depending on the ratio of the two standard deviations and the magnitude of $r_{X_1\Delta X}$.

Some authors (for example, Markus, 1980) seem to imply that regression to the mean is *always* present in change models. This is not the case. However, there are at least two reasons why regression to the mean occurs with great frequency. First, many phenomena in the real world have relatively stable variances. Whenever this is so, regression to the mean will necessarily exist. To see why this is so, consider the fact that the variance will increase if an individual's initial score on some variable is positively related to subsequent change, *or* (assuming that some individual-level change has occurred) if initial scores are totally unrelated to change. Any change at the individual level will necessarily add to the initial variance of the variable being studied. Only a negative correlation between initial scores and change can compensate this added variance to yield an over-all variance that is constant over time.

So far we have mentioned regression to the mean as a substantive phenomenon; however, it can also arise from measurement error. We will have more to say about this type of regression effect in Chapter 10.

An Illustration: The Stability of Alienation

To illustrate the sorts of algebraic manipulations we have been discussing, we examine the results of a three-wave panel study of a population in a rural region of Illinois in the years 1966, 1967, and 1971 (Wheaton *et al.*, 1977). Interest centers on the stability of alienation and the causal impact of socioeconomic status (SES) on change in alienation.

In their analysis of the data, the authors of the study convincingly demonstrated the value of using complex measurement models to describe the relationships between true and measured indicators of alienation and SES. Figure 2.2 shows the full model of the relationship between true and observed alienation and SES estimated in their work. Measured indicators are in lower case letters, and unmeasured true scores are in upper case letters. The two indicators of SES are ses_{11} and ses_{12}; the two indicators of alienation are a_{1t} and a_{2t}, each measured at time points $t = 1$ (1966), 2 (1967), and 3 (1971). The symbol e represents measurement error in the indicator variables, while U represents the disturbance of a true score after it has been purged of measurement error.

We will be concerned here only with the true scores. These were not directly measured, but were inferred from the pattern of relationships among the observed indicators. The part of the model that involves the true scores is just-identified, so that we can study change and its determinants in this part of the model. The correlation matrix for the true scores is given in Table 2.2.

Table 2.2

Correlations (Lower Half) and Variances (Diagonal) of True Scores in Figure 2.3

	SES	A_1	A_2	A_3	U_2	U_1
SES	1.000					
A_1	−.526	1.000				
A_2	−.558	.808	.836			
A_3	−.620	.630	.701	.900		
U_2	−.151	.287	.232	.809	.476	
U_1	.000	.000	.000	.000	.000	.270

We shall have nothing to say about the estimation procedure used in the original analysis. The parameter estimates obtained in that analysis for the part of the model of concern to us are presented in Figure 2.2, with standardized coefficients given in parentheses.

Before presenting our reanalysis, we note some of the conclusions the authors draw from these estimates. They conclude that alienation is fairly stable over 1 year (times 1 to 2) and only about half as stable over 4

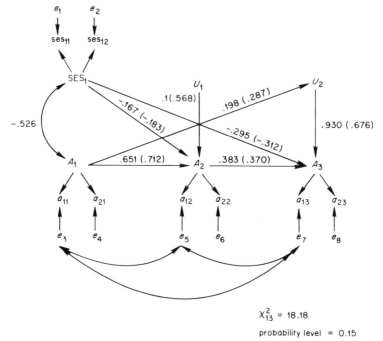

$$\chi^2_{13} = 18.18$$

probability level = 0.15

Figure 2.2. Complex measurement model of the influence of SES on alienation. Source: Figure 6 in Wheaton et al. (1977, p. 122).

years (times 2 to 3); that alienation has a positive impact on subsequent alienation; and that initial SES has a negative impact on subsequent alienation. Another point of interest, not discussed by the authors, is that the direct effect of initial alienation (standardized) is about four times that of SES over 1 year, and falls to little more than equal importance over 4 years. The reanalysis will show that three of these four conclusions must be modified when our decomposition procedure is employed.

The importance of each predictor variable on changes in alienation over the intervals 1–2 and 2–3 can be derived from Figure 2.2 and Table 2.2 using Eqs. (2.7a) and (2.7b). For example, once we have used Eq. (2.8) to compute $s_{\Delta A_{2-1}} = .598$, we can proceed to calculate

$$\beta_{\Delta A_{2-1}A_1.SES} = (b_{A_2A_1} - 1)(s_{A_1}/s_{\Delta A_{2-1}}) = (.651 - 1)(1/.598) = -.583$$

and

$$\beta_{\Delta A_{2-1}SES.A_1} = b_{A_2SES.A_1}(s_{SES}/s_{\Delta A_{2-1}}) = (-.167)(1/.598) = -.279.$$

Similarly, by using Eqs. (2.12a) and (2.12b), we find

$$\beta_{A_2A_1.\Delta A_{2-1}} = s_{A_1}/s_{A_2} = 1/.915 = 1.094$$

and

$$\beta_{A_2\Delta A_{2-1}.A_1} = s_{\Delta A_{2-1}}/s_{A_2} = .598/.915 = .654.$$

The results of these calculations are displayed in Figure 2.3. The figure makes explicit the impact of A_t on A_{t+1} through its direct and indirect components. The estimated stability coefficients in Figure 2.2 can be reproduced by summing the direct and indirect paths from A_t to A_{t+1}. For example, the stability coefficient between A_1 and A_2 in Figure 2.2 is determined from Figure 2.3 as $1.094 + (-.583)(.654) = .712$.

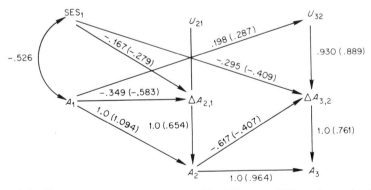

Figure 2.3. True score parameter estimates of the influence of SES on change in alienation over time.

Comparing the impacts of earlier alienation over the two intervals on the change components of subsequent alienation, we find that A_1 has less impact on $\Delta_{A_{2-1}}$ than A_2 does on ΔA_{3-2}. Even though initial alienation does not exhibit a strong impact on change in alienation over an interval of one year $(-.349)$, the impact of time 2 alienation is quite strong $(-.617)$ over the longer interval of four years. This is true even though the overall stability coefficient of A_2 on A_1 $(.712)$ in Figure 2.2 is larger than the comparable coefficient for the regression of A_3 on A_2 $(.370)$. This finding is unexpected since the time interval 2–3 is four times as long as the time interval 1–2.

Of even greater interest, the effects of initial alienation on change are negative in both intervals. Thus these causal impacts lessen the initial differences in alienation among members of the sample. The negative causal impacts, although consistent, operate on changes that are a good deal smaller parts of the later score variances than are the stability components. As a result, overall regression estimates of stability are positive. This corresponds to case D in Appendix A (the case where $r_{X_1 \Delta X}$ is negative, but smaller in magnitude than $s_{X_1}/s_{\Delta X}$). In this case, overall regression estimates of stability are positive.

Finally, by examining the impact of SES on the change component of alienation on the basis of the parameters in Figure 2.3, one can see a good deal more consistency than is found in Figure 2.2. In particular, the direct effect of SES on change in alienation is consistently about 50% as large as that of initial alienation on change. This consistency is masked in Figure 2.2, which confounds the differential lack of change across the two intervals with the consistent impact of SES on change.

In all, the decomposition of alienation into change and lack of change components has uncovered a number of patterns missed when the static score equations were interpreted directly. In particular, the negative impacts of initial alienation on subsequent changes, the greater impact of initial alienation on change over the longer rather than the shorter time period, and the consistency in the relative effects of alienation and SES on change are all missed when the decomposition is not carried out.

Most of these insights can be gleaned from the data without complicated computation. It is very easy to compute change variances from Eq. (2.8). A comparison of these to the static score variances will give the researcher an idea of the relative change across two or more time intervals. Then a mental subtraction of 1 from the regression coefficients $b_{A_3A_2}$ and $b_{A_2A_1}$ will show the influence of initial alienation on these change components. A mental comparison of the relative ratios of slopes from later alienation to earlier alienation and SES after subtracting the constant from the autoregression will show the relative contributions of the two predictors to change. Only when the analyst is interested in standardized coefficients between static scores and change scores is it necessary to carry out tedious calculations for the reparametrization.

Even though most of the interpretation of panel results in terms of change components can be done mentally or with the help of simple computations, this illustrative reanalysis shows that they can help the researcher come to a deeper understanding of the data than is possible when the results are interpreted in terms of static scores alone.

Appendix A: An Analytic Investigation of the Stability Coefficient

The researcher interpreting the magnitude of a stability coefficient should be aware of the various influences that go into creating this coefficient. We can make use of Eq. (2.2) to provide us with some insights into these influences.

Let us begin by substituting Eq. (2.2) into the equation for a gross stability, or product-moment correlation, coefficient. This gives us

$$r_{X_1 X_2} = \frac{s_{X_1} + r_{X_1 \Delta X} s_{\Delta X}}{(s_{X_1}^2 + s_{\Delta X}^2 + 2r_{X_1 \Delta X} s_{X_1} s_{\Delta X})^{1/2}} \tag{2A.1}$$

which expresses the overall correlation between X_1 and X_2 in terms of two parameters—the correlation between X_1 and ΔX, and the ratio of $s_{\Delta X}$ to s_{X_1}. In considering this dependence, five combinations of these two quantities are of interest.

1. When $r_{X_1 \Delta X} = 1$, $r_{X_1 X_2}$ will also equal 1, regardless of the ratio $s_{X_1}/s_{\Delta X}$. This corresponds to the case when change is completely determined by the initial score.

2. When $r_{X_1 \Delta X}$ is positive, but less than 1, $r_{X_1 X_2}$ will remain positive, but will be larger in magnitude than $r_{X_1 \Delta X}$.[7] Since X_1 is perfectly correlated with the stable part of X_2 (which is just X_1) and imperfectly correlated with the change part of X_2, namely ΔX, $r_{X_1 X_2}$ will lie between 1 and $r_{X_1 \Delta X}$, the exact magnitude depending on the ratio of the two standard deviations.

3. When $r_{X_1 \Delta X} = 0$, Eq. (A2.1) reduces to

$$r_{X_1 X_2} = s_{X_1}/(s_{X_1}^2 + s_{\Delta X}^2), \tag{2A.2}$$

---

[7] It is not necessary that $r_{X_1 \Delta X}$ be positive for $r_{X_1 X}$ to exceed $r_{X_1 \Delta X}$. The necessary condition is that $r_{X_1 \Delta X} > -s_{X_1}/2s_{X_2}$.

which represents the fraction of $s_{X_2}^2$ directly due to X_1 being a part of X_2. Since X_1 has no impact on ΔX, the magnitude of the coefficient depends only on the magnitude of the variance of the change score relative to the time 1 variance. When little change takes place over the interval, X_1 will make up a large part of X_2 and the coefficient will be large; when a great deal of change takes place, the coefficient will be small. The correlation $r_{X_1X_2}$ will always be positive, and in all but the most extreme cases it will be significantly different from zero even though X_1 has no causal impact on change.

4. When $r_{X_1\Delta X}$ is *negative* but smaller in magnitude than $s_{X_1}/s_{\Delta X}$, $r_{X_1X_2}$ will remain *positive,* and can be very different in magnitude from $r_{X_1\Delta X}$. An example of this can be seen in the study of body weight over time. In general, overweight people tend to lose weight over time, underweight people tend to gain, and people of average weight tend to remain stable. This effect is the substantive type of "regression to the mean" we mentioned in the text of Chapter 2. It is a phenomenon well-known in psychological testing.

As long as s_{X_1} exceeds $s_{\Delta X}$, $r_{X_1X_2}$ will become *more positive* as $r_{X_1\Delta X}$ becomes *more negative,* reaching $+1$ when $r_{X_1X_2}$ is -1. This represents a perfect dynamic equilibrium: those with initially high scores will decrease more (or increase less) than persons with initially low scores. Since, in the limiting case, the pattern of change is perfectly determined by the initial scores, the correlation between initial and final scores is 1. Where the variance of time 2 scores remain the same as the variance of time 1 scores as well, the aggregate characteristics of a normally distributed population will remain entirely unchanged between time 1 and time 2.

When $s_{X_1} = s_{\Delta X}$, $r_{X_1X_1}$ will approach 0 from above as $r_{X_1\Delta X}$ approaches -1. This means that a low value of $r_{X_1X_2}$ can be masking a strong negative causal impact of the X_1 score on the change component of X_2. For example, when $r_{X_1\Delta X}$ is large and negative $(-.9)$ and $s_{X_1X_2} = s_{\Delta X}$, $r_{X_1X_2}$ will be small and positive (.22). For the case $s_{X_1} < s_{\Delta X}$ the masking effect will be less severe, but can still be substantial.

5. Finally, when $r_{X_1\Delta X} < -s_1/s_{\Delta X}$, $r_{X_1X_2}$ will also be negative, and will approach -1 as $r_{X_1\Delta X}$ approaches -1.

These five cases represent all the patterns of relative influence one might encounter in the interpretation of a stability coefficient. In the limiting case where $r_{X_1\Delta X}$ is equal to 1, the interpretation of stability is simple. However, when $r_{X_1\Delta X}$ is less than 1, confusion in the interpretation of $r_{X_1X_2}$ can arise because the relative importance of lack of change and the causal impact of X_1 on change are not distinguished. This confusion is expecially likely to occur when the variance of ΔX approaches or exceeds the variance of X_1. In that case, a small and positive value of $r_{X_1X_2}$ can mask a strong, negative causal impact of X_1 on ΔX. Since it is common in survey research to find that the covariance of X_1 with ΔX is negative once corrections for measurement error have been made, and since the contribution of ΔX to X_2 will generally increase as the interval between observations increases, such masking effects should be expected in panel research conducted over long time intervals.

Appendix B: Brief Introduction to LISREL

At several points in the text we mention the computer program LISREL. This program was developed by Karl Jöreskog at the University of Uppsala in Sweden, and is currently available in the United States through National Education Resources, Inc., in Chicago.

Since several quite accessible introductions have appeared recently in the literature (Jöreskog, 1973; Jöreskog and Sörbom, 1975, 1977, 1979; Long, 1976; Wheaton *et al.* 1977), we do not describe the specifics of LISREL here, but provide a few words of introduction for those readers who are not already familiar with the type of problems LISREL is designed to handle.

LISREL provides the researcher with a means of efficiently arriving at parameter estimates in over-identified linear models. It operates on a covariance matrix of observed scores and iteratively derives parameter estimates for a specified model in such a way that the discrepancy between the observed covariance matrix and the covariance matrix predicted by the model is minimized. In other words, LISREL iteratively adjusts the parameter estimates in the model so as to optimize the overall fit of a postulated structure and the observed patterns of relationships in the data. The overall adequacy of fit of any given model can be tested with a chi-square test of statistical significance, and the relative fits of two or more models can be compared.

When a model is just-identified, the LISREL solution converges on an ordinary least squares solution. When a model is over-identified and the observed variables are adequately described by their first two moments (means, standard deviations and correlations), the LISREL solution is maximum-likelihood.

LISREL estimation can be used for any type of linear structural equation model as long as the equations expressing the observed variances and covariances in terms of the model parameters are nonsingular. In particular, the program can handle simultaneous equations for the reciprocal influence of jointly dependent endogenous variables, and can estimate correlations among errors. It can do exploratory and confirmatory factor analysis.

The currently available version of LISREL (which as of this writing is LISREL IV) permits the researcher to constrain estimates of coefficients to be equal, an option we mention in Chapters 3, 6, and 9. It also permits parameters to be fixed to given numerical values, an option discussed in connection with the use of generalized least squares to deal with serially correlated errors. In addition, it allows measurement error models that distinguish between observed and latent variables to be estimated. Models of this kind are discussed in Chapter 10.

3

The Causal Analysis of Change

In the preceding chapter we considered the conceptualization and measurement of change and its determinants. Although we invoked the concept of causality in our discussion, we did not ask how one might determine empirically that one variable is the cause of another.

A causal relationship between X and Y implies, at a minimum, an association of some sort between X and Y, lack of spuriousness in this association, and evidence about the direction of causality. The first criterion alone is not sufficient. Indeed, it is a truism in the study of elementary statistics that correlation need not imply causation. A statistical correlation or association between X and Y may mean that X causes Y, but it may also mean that (a) Y causes X, (b) X and Y are each causes of one another, (c) X and Y are both caused by other variables, or (d) there is error in the measurement of X and Y, and measurement errors for these two variables are correlated.

These possibilities are by no means mutually exclusive; indeed, several of them may be true at the same time in a given set of data. Assuming the existence of a correlation between X and Y (a simple empirical question), the fundamental problem in establishing that X is a cause of Y is to exclude alternative explanations of the correlation between the two variables. Or, when several of these mechanisms are present simultaneously, the goal will be to estimate the contribution that each makes to the observed correlation.

The way lack of spuriousness and evidence concerning the direction of causality are handled depends on the way the data have been collected. In

25

experimental research, spuriousness and direction of causality are evaluated by randomization and manipulation, respectively. The random assignment of subjects to a treatment or control group guarantees that there are no differences between the two groups on any variable apart from those due to statistical fluctuation, and probability theory can tell us the likelihood of a fluctuation of given magnitude. Possibility (c), that X (the treatment) and Y (outcome) are both caused by a third variable Z can thus be excluded, as randomization destroys any possible relationship between X and Z (Z and Y may remain correlated even after randomization, but this is irrelevant to the question of the effect that X has on Y as long as Z has no effect on Y). Similarly, the randomization establishes that initial differences in the criterion (Y) have no effect on X (that is, do not influence whether someone is placed in a treatment or control group). This eliminates possibilities (a) and (b). Since assignment to the two groups is presumably under the control of the experimenter, the question of measurement error in X should not arise.[1]

For better or worse, many of the social processes we wish to investigate cannot be studied in this way. The experimental design is frequently impractical, unethical, or too divorced from reality to shed light on the issues of concern to us. We are thus forced to draw inferences from nonexperimental data. Under this circumstance, causal analysis becomes much more difficult. Though not a complete substitute for the experimental design, panel analysis goes farther toward resolving the ambiguities in causal inference than other forms of analysis. To see why this is so, we review briefly the difficulties of drawing causal inferences from cross-sectional data, and then present an analogous discussion for panel data.

Causality in Cross-Sectional Analysis

With cross-sectional data, we would determine the linear effect of X on Y by regressing Y on X. If additional variables Z are believed to cause both X and Y, they are introduced into the regression as well. We write

$$Y = bX + cZ + u \qquad (3.1)$$

where u is a disturbance term, b and c are unstandardized regression coefficients, and all variables are measured from their means.

Note that *all* variables that influence both X and Y must now be intro-

---

[1] For a discussion of the complications that can arise in experimental research when treatment outcome can influence the treatment itself, see Miller (1971).

duced into the regression equation *explicitly,* for estimates of regression coefficients may be biased if any are omitted. In the experimental design, no specific attention had to be given to these variables.

An additional complication concerns the question of causal order. Suppose that in reality Y is a cause of X, but X has no effect on Y. We can still regress Y on X and Z and obtain an estimate for b, but the interpretation of this coefficient as representing the strength of the influence of X on Y would be totally wrong. And there would be no way to know this from the computations themselves.

When the possibilities that X causes Y and Y causes X are both tenable, one can write a set of simultaneous structural equations to express this mutual dependence. Here one includes as predictors of X all those variables believed to cause X (including Y), and as predictors of Y all those variables believed to cause Y (including X). With cross-sectional data, however, these equations can be solved only if extremely stringent assumptions are made about the effects of exogenous variables on X and Y. In particular, if the effect of X on Y is to be estimated, the researcher must specify in advance the numerical magnitude of the effect that at least one exogenous variable has on Y. This is usually done by assuming that a given exogenous variable has no effect on Y. Where assumptions of this kind can be justified by substantive considerations, these methods will permit the various contributions to the correlation between X and Y to be estimated (Rao and Miller, 1971; Johnston, 1972; Namboodiri, Carter, and Blalock, 1975; Hanushek and Jackson, 1978). Too often for comfort, though, social science theory is incapable of justifying the assumptions needed to estimate these equations. Under these circumstances, a statistical analysis of cross-sectional data based on the assumption that causal influences are at work in one direction only would run a serious risk of bias in estimating the causal influence assumed to be present. But the more sophisticated analyses that abandon the assumption of unidirectional causality cannot be carried out.

We emphasize that problems where this difficulty arises are not esoteric, or confined to some tiny area of specialized interest. The reader will have no difficulty in thinking of a multitude of examples from virtually any field in the social sciences.

Intuitively, it seems clear that one can guard against the risk of bias originating in cross-sectional analyses from erroneous assumptions about the causal structure linking the variables by looking at the effect of one variable on the *change* in the other variable. If X has no causal effect on Y, then Y will not change between two observations even if X changes, and no mistake will be made. Or we can draw on the equivalence between the analysis of change scores and the analysis of static scores demonstrated in the previous chapter, and regress Y_2 on X_2, controlling for X_1. Evidently, this requires panel data.

The remainder of this chapter will be concerned with various approaches to the evaluation of direction of causality using panel data. A good deal of confusion exists about the use of panel data to make this type of evaluation. It is true that causal priority between any two variables can be inferred on the basis of panel data. But it is not sufficiently appreciated that assumptions about causal associations must be made before inferences of this sort can be made. These assumptions are no less central to panel analysis than to cross-sectional analysis. In neither can we *demonstrate* causality on the basis of empirical information alone. This is something that can only be done on the basis of experimental manipulation. Instead we are forced to *infer* causal direction from parameter estimates within the context of an assumed causal model. The advantage of a panel analysis over a cross-sectional analysis, then, will not lie in our being relieved of the necessity to make *some* causal assumptions, but in the possibility of making weaker assumptions than are required with cross-sectional data.

A General Two-Wave, Two-Variable Model

We begin our discussion of causal inference in panel models by analyzing the most general linear model in which two variables X_t and Y_t are measured at times 1 and 2. Two-wave models arise often in practice, and at the same time are simple enough to display the features of interest. Assuming perfect measurement, ignoring the possibility of any unmeasured common causes of X and Y, and expressing X and Y as deviations from their means, the causal relationships among these variables are

$$X_2 = b_1X_1 + b_2Y_1 + b_3Y_2 + u, \qquad (3.2a)$$

$$Y_2 = d_1Y_1 + d_2X_1 + d_3X_2 + v, \qquad (3.2b)$$

where u and v are, respectively, the residuals of X_2 and Y_2. The path diagram corresponding to this general model is shown in Figure 3.1.

The six correlations among X_1, X_2, Y_1, and Y_2 can be expressed in terms of the parameters of this model in the following manner:

$$r_{X_1Y_1} = \text{given (i.e., observed)}, \qquad (3.3a)$$

$$r_{X_1X_2} = b_1 + b_2r_{X_1Y_1} + b_3r_{X_1Y_2}, \qquad (3.3b)$$

$$r_{X_1Y_2} = d_1r_{X_1Y_1} + d_2 + d_3r_{X_1X_2}, \qquad (3.3c)$$

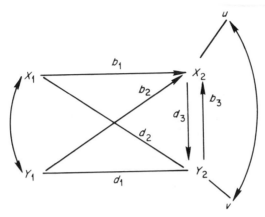

Figure 3.1. A general two-wave, two-variable panel model.

$$r_{Y_1Y_2} = d_1 + d_2r_{X_1Y_1} + d_3r_{Y_1X_2}, \tag{3.3d}$$

$$r_{Y_1X_2} = b_1r_{X_1Y_1} + b_2 + b_3r_{Y_1Y_2}, \tag{3.3e}$$

$$r_{X_2Y_2} = b_1r_{X_1Y_2} + b_2r_{Y_1Y_2} + r_{Y_2u}. \tag{3.3f}$$

There are eight unknown parameters in these six equations (the three b_i and three d_j, the time 1 cross-sectional correlation $r_{X_1Y_1}$, and the correlation $r_{Y_2}u$). It is impossible to solve uniquely for any of these unknowns except $r_{X_1Y_1}$ by manipulating the six equations (Duncan, 1969a). To solve for b_1, b_2, b_3, d_1, d_2, and d_3 we must make some restricting assumptions. This is what we meant when we said that causal direction in panel analysis (or for that matter, in any nonexperimental data analysis) must be inferred on the basis of a model that imposes some restrictions on a completely general causal structure. The availability of panel data does not entirely eliminate the necessity of making some assumption about causal order.

When only the four variables X_t, Y_t ($t = 1$, 2) are considered, two approaches can be taken to the imposition of restrictions on the solution. In the first, we assume that the values of one or more parameters are fixed at specified values, usually zero. In the second, we assume that the relative magnitude of two parameters is some fixed value. When additional variables are introduced as controls in the model, a third approach is possible: to impose some constraints on the associations between these variables and the four variables of central interest.

In the remainder of this chapter we examine how these approaches have been used to identify a restricted version of the two-wave, two-variable model in several pieces of empirical research. Then we examine causal inference in the multiwave panel model.

Fixing Values of Parameters

The most common way to impose restrictions on the general two-wave, two-variable model is to fix the values of some of the parameters in the model. This approach is illustrated by the recent work of Wheaton (1978).

The findings from community surveys are quite consistent in showing that socioeconomic status is inversely related to feelings of psychological distress (Dohrenwend and Dohrenwend, 1969). Confusion exists, though, as to the direction of causal influence between these two variables (Liem and Liem, 1978). On the one hand, there is reason to believe that social class position is related to stressful life experiences and that the latter directly cause subjective feelings of distress (Myers *et al.*, 1974). On the other hand, it has also been argued that an individual's level of emotional functioning, one indicator of which is his or her subjectively experienced sense of well-being, can importantly determine socioeconomic achievement (Turner and Wagonfeld, 1967; Turner, 1968).

Wheaton (1978) studied this controversy by conceptualizing it as an issue of contending causal hypotheses in a two-wave, two-variable panel model. Using a community sample of individuals interviewed in 1965 and again in 1971, and controlling for the socioeconomic achievements of respondents' parents and for the achievements of respondents themselves as of 1960, he estimated a cross-lagged version of the model in Figure 3.1 for indices of socioeconomic status and psychological distress in 1965 and 1971. In terms of the notation used here, Wheaton assumed that $b_3 = d_3 = 0$. This implies that current socioeconomic status has no influence on current feelings of distress (a highly questionable assumption) and that current feelings of distress have no influence on current socioeconomic status.

On the basis of these assumptions, the following estimates were obtained, expressed here in standardized form:

$$PD_2 = .681\ PD_1 - .010\ SES_1,$$
$$SES_2 = -.001\ PD_1 + .928\ SES_1,$$

where PD = psychological disorder and SES = socioeconomic status. Both variables are highly stable, SES almost perfectly so. Neither variable significantly influences the other.

The assumptions Wheaton made to identify these parameters are not the only ones he could have made, and one of these is seemingly an unreasonable one: that current socioeconomic status has no influence on psychological functioning. A much more reasonable constraint would be that past socioeconomic status has no influence on current psychological functioning once current SES is controlled. This means that we might estimate a lagged effect of distress on socioeconomic status and a contemporaneous effect of SES on distress. If we identify X in Figure 3.1 with class and Y with distress, then these constraints amount to the assumption that d_2 and b_3 are zero.

In the data analyzed by Wheaton, a recomputation of the model with the assumption $d_2 = 0$ substituted for the assumption $d_3 = 0$ leaves the conclusions drawn in the original analysis unchanged. This is so because SES is almost perfectly stable over the 6-year interval, making the choice between a lagged and contemporaneous causal effect unimportant. However, this fortunate state of affairs will not always prevail. The researcher should consequently base the choice of identifying constraints on a solid substantive rationale. We turn now to a study in which this detailed consideration of constraining options was made.

Constraining Associations with Control Variables

A much more flexible approach to constraining one's model entails the use of control variables. This approach is illustrated in the recent work of Kohn and Schooler (1978), who were concerned with the mutual causal influences between the character of people's jobs and their intellectual functioning. In 1964 and again in 1974 panel data were collected on a sample of 667 working males, and measures were obtained about the substantive complexity of their jobs (SC) and their degrees of intellectual flexibility (IF).

Although the authors recognized explicitly that it would be necessary to make some identifying assumptions to solve for specific causal coefficients, they felt there was good reason to expect both cross-lagged and cross-contemporaneous effects between these two variables. Therefore, they rejected the approach to identification based on making restrictive assumptions about two or more of these parameters.

Instead, a series of assumptions were made about the effects of certain control variables on the time 2 (1974) measures of IF and SC. As the authors noted,

> for the model to be adequately identified, the direct effects of one or more exogenous variables on 1974 intellectual flexibility must be assumed to be zero; similarly, the direct effects of one or more exogenous variables on 1974 substantive complexity must also be assumed to be zero.

In fact, as noted, the solution can be identified by assigning *any* fixed nonzero value to the direct effect of an exogenous variable on the reciprocally dependent endogenous variables. In practice, though, it is usually much easier to find a predictor variable for which a plausible case can be made that its effect on the criterion is zero than to come up with one for which a nonzero effect of given magnitude can plausibly be assumed.

The assumption Kohn and Schooler made about 1974 substantive complexity was based on the following reasoning:

> We posit that background characteristics that would not be interpreted as job credentials by employers (even by discriminatory employers) do not directly affect the substantive complexity of the 1974 job; these variables are thus used as instruments to identify the equation. The rationale is that these variables—maternal and paternal education, paternal occupational level, maternal and paternal grandfathers' occupational levels, urbanicity and region of origin, and number of children in the parental family—may very well have affected men's job placement earlier in their careers. By the time that men are at least 10 years into their careers, however, these variables should no longer have any direct effect on the substantive complexity of their jobs, certainly not when the substantive complexity of their 1964 and earlier jobs are statistically controlled [pp. 41–42].

The authors were unwilling to make a similar set of assumptions about the effects of background characteristics on 1974 intellectual flexibility. Since they had available a measure of the substantive complexity of the jobs respondents held prior to 1964, they made the following assumption about 1974 intellectual flexibility instead:

> We posit that the substantive complexity of earlier jobs should have no direct effect on the men's intellectual flexibility in 1974, when the substantive complexity of their 1964 and 1974 jobs are statistically controlled [p. 42].

The final model estimated by Kohn and Schooler is reproduced in Figure 3.2. The critical identifying assumptions are clearly displayed. The results are equally clear in showing that intellectual flexibility and substantive complexity reciprocally cause each other, but at different lags. First, the effect of intellectual flexibility on the substantive complexity of jobs is lagged and substantial, approximately equal to the effect of the earlier substantive complexity of work in 1964 on the substantive complexity of the job held in 1974. Thus people who are intellectually flexible do seek out more complex jobs. This tendency has a major effect on the complexity of the jobs people hold, though the complexity of their earlier jobs is an additional important influence. Second, the effect of substantive complexity on intellectual flexibility, although modest in comparison with the effect of IF_1 on SC_2, is statistically significant and instantaneous rather than lagged. To some extent, the substantive complexity of people's current jobs appears to influence their intellectual flexibility. This effect, however, is much smaller than might be inferred from a cross-sectional analysis carried out on the basis of the assumption that SC influences IF, but IF has no effect on SC.

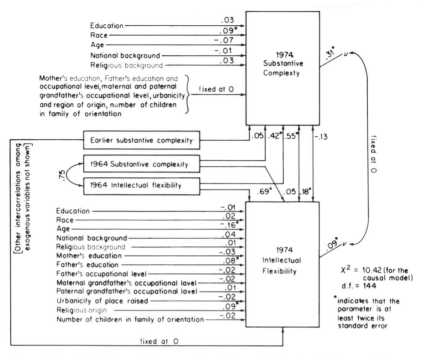

Figure 3.2. The final model estimated by Kohn and Schooler (1978, p. 40). This was Figure 4 in the original source, with the heading "Reciprocal effects of substantive complexity and intellectual flexibility: full model; coefficients shown are standardized."

The identifying assumptions made here were well-considered and justified. Nonetheless, Kohn and Schooler were concerned about the possibility that results of their analysis might be sensitive to these particular assumptions. To evaluate this possibility they carried out a sensitivity analysis of their model. This analysis showed that their parameter estimates were unchanged when somewhat different constraints were used to identify the model. We will discuss sensitivity analyses in panel models in Chapter 9.

Consistency Constraints in Multiwave Models

At times one may wish to estimate both cross-lagged and contemporaneous effects between two variables, but will know of no control variables for which identifying parameter constraints can plausibly be assumed. Under this circumstance, two waves of data will not be sufficient to permit model parameters to be estimated, and panel analysis will provide no advantage over a cross-sectional analysis.[2] But if *three or more* waves of data are collected, it may still be possible to identify one's model by imposing consistency constraints on the parameters. This approach to identification is illustrated in the recent work of Greenberg, Kessler, and Logan (1979). This analysis was concerned with the impact that marginal changes in the chances of being arrested following a crime have on crime rates.

By virtue of their possible deterrent and restraining effects, it would seem likely that arrests reduce crime rates. Estimation of the size of this effect is complicated by the posibility that police resources, which are fixed in the long run, could be overwhelmed when crime rates increase. If that occurs, higher crime rates would lead to a decline in the proportion of crimes that result in an arrest. On the other hand, high crime rates might tend to enhance police efficiency by increasing public support for higher police budgets and more effective deployment strategies. If, indeed, the relationship between arrests and crime rates is a reciprocal one, a conventional multiple regression analysis would yield biased estimates of these effects, just as in our previous examples.

An additional complication is that the causal influences at work may be instantaneous as well as lagged.[3] Little delay is expected in the deterrent effect of an arrest, since arrests receive publicity mainly within a few days of their occurrence. Saturation-of-resources effects would likewise be expected to be primarily a short-run phenomenon. On the other hand, public recognition that crime rates have risen may not occur immediately, and further delay would be expected before the public's demand that "something be done" about crime results in larger appropriations or innovations in policing methods.

Some researchers have tried to take account of the possible instantaneous reciprocal influence of crime rates and criminal justice sanctions in cross-sectional analyses through the use of simultaneous equation methods of the sort described earlier, but two considerations make this approach less than ideal. First, as we pointed out earlier in this chapter, in discussing cross-sectional analyses, the regression coefficients of some outside predictor variables with arrest and crime rates must be fixed a

[2] Should the researcher be willing to fix the ratio of lagged and contemporaneous effect at some specific value, then the two wave model *is* identified.

[3] The word "instantaneous" should not be interpreted too literally. As long as an effect occurs primarily within a time span that is short by comparison with the time interval between observations, we will refer to it as instantaneous.

priori at some prespecified values. Kohn and Schooler employed this sort of constraint in their analysis. As we saw in our consideration of their reasoning, there are times when quite plausible cases can be made for constraints such as these. Yet in the case of crime and arrest, too little is currently known to regard such assumptions with confidence. Indeed, Nagin (1978) and Fisher and Nagin (1978) have argued on the basis of an extensive literature review that the constraints of this sort used in much of the published research on crime deterrence are implausible on substantive grounds. Simultaneous equation methods are highly sensitive to bias due to misspecifications of this sort. Should the restrictions imposed to achieve identification be in error, then, the estimates of the mutual effects of crime and arrest on one another may be biased, perhaps quite badly.[4]

A second difficulty is that in these cross-sectional analyses, lagged effects are absorbed into later measures of enforcement and crime. Yet if the saturation effect of crime on resources tends to reduce police effectiveness, while public alarm over rising crime rates tends to enhance this effectiveness, the instantaneous and lagged effects of crime on police efficiency will be of opposite sign. The two effects will tend to neutralize one another, and thus lead to an underestimation of the influence of crime on enforcement. To avoid this possible source of bias, a model of the effects of crime on enforcement must include both short- and long-term effects.

It is obvious that panel models are required here. In fact, several investigations of crime rates have utilized such models involving two waves of data (Tittle and Rowe, 1974; Logan, 1975; Pontell, 1978). However, in all of these analyses identification was achieved by assuming that either lagged or contemporaneous effects were absent—assumptions which the considerations we have discussed above call into question.[5] Yet, at the same time, the arguments of Nagin (1978) and Fisher and Nagin (1978) suggest that the solution adopted by Kohn and Schooler, of using outside predictor variables to make the necessary identification constraints, will be unsatisfactory. If this is true, then the problem is intractable with only two waves of data.

Under some conditions this problem can be overcome through the use

---

[4] This potential source of bias is equally present in time series analysis.

[5] For example, in an analysis of city crime and arrest rates in Florida for 1971 and 1972, Tittle and Rowe (1974) estimated two-wave panel models assuming that cross-instantaneous effects vanish. Logan's (1975) analysis of state crime and arrest data was carried out on the basis of several sets of specifications. In one, lagged effects were assumed to vanish; in a second, the lagged effect of crime on arrest and instantaneous effect of arrest on crime were assumed to vanish; and in a third, crime was assumed to affect arrest only instantaneously, and the disturbance terms were assumed to be uncorrelated. On substantive grounds these assumptions seem questionable.

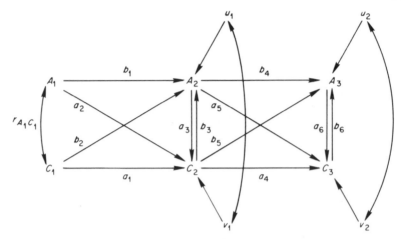

Figure 3.3. A three-wave, two-variable panel model for crime rates and clearance rates.

of multiwave (three or more wave) panel models. This is done by making assumptions about the consistency of effects rather than about their values. These assumptions are far weaker, and hence more plausible, than the identification restrictions we have considered thus far. As the consistency approach permits the data to play a larger role in the analysis, the approach seems particularly suitable when theory is not well developed. The study of crime causation is a prime example of such a situation.

We illustrate the consistency procedure by considering the model for the crime rate C and the arrest clearance rate (ratio of arrests to crimes) A shown in Figure 3.3. We assume that the causal effects are reciprocal, and that they are adequately expressed by instantaneous terms, and terms lagged by a single time unit.[6] The structural equations for this model are

$$C_2 = a_1 C_1 + a_2 A_1 + a_3 A_2 + v_1, \qquad (3.4a)$$

$$C_3 = a_4 C_2 + a_5 A_2 + a_6 A_3 + v_2, \qquad (3.4b)$$

$$A_2 = b_1 A_1 + b_2 C_1 + b_3 C_2 + u_1, \qquad (3.4c)$$

$$A_3 = b_4 A_2 + b_5 C_2 + b_6 C_3 + u_3. \qquad (3.4d)$$

As done previously, the correlation between time 1 variables is taken into account, but not subjected to causal analysis. We permit the error terms u

---

[6] We assume that longer lagged effects are not present only to simplify the discussion. Where there is reason to believe they exist, they can be incorporated in the analysis without difficulty.

and v to be cross-sectionally correlated, but for the moment we assume that serial correlation among the errors is not present[7] (that is, we assume that $r_{u_1u_2} = r_{v_1v_2} = 0$).

Concentrating on the equations for C_2 and C_3, we note that there are six regression coefficients to be estimated. By taking covariances of Eq. (3.4a) with C_1 and then with A_1 (that is, by using C_1 and A_1 as instrumental variables), we obtain two normal equations that can be used to help estimate these parameters. By repeating the procedure for Eq. (3.5b) using C_2 and A_2 as instruments, we obtain an additional two normal equations.

We cannot similarly use A_3 to identify Eq. (3.4b) because the reciprocal influence between A_3 and C_3 implies that A_3 cannot have a vanishing correlation with v_2.[8] For the same reason, neither A_2 nor A_3 can be used to help identify Eq. (3.4a).

Seemingly, we should be able to derive an additional two normal equations by taking the covariances of Eq. (3.5b) with C_1 and A_1, as these variables are not correlated with v_2. Indeed, one can do this, and the equations so obtained are valid. The trouble is that they contain no new information beyond that already contained in the other four normal equations. To see why this is so, note that Eqs. (3.4a) and (3.4c) can be solved to yield "reduced form" equations that express C_2 and A_2 in terms of C_1, A_1 and the disturbance terms u_1 and v_1. It follows that the normal equations derived by using C_1 and A_1 as instruments will be linear combinations of those derived by using C_2 and A_2 as instruments.

With six parameters and four normal equations that can be used to estimate them, two additional pieces of information are needed if a unique solution is to be found. In this respect, we are in the same position as when we considered the two-wave, two-variable model. However, we now have more options for identifying the model than we did in the two-wave case. We could, of course, assume that two of the parameters have given values, but it would be difficult to justify these assumptions. Instead of assuming that certain parameters have known numerical values, we can assume that certain effects remain *constant* over the period between the first and third waves, a fairly weak assumption.[9] In the notation of the

---

[7] The implications of nonvanishing serial correlation among the error terms will be discussed in Chapter 9. In the empirical example we review, however, we do include serial correlations in the model.

[8] Note that in Eq. (3.4d), A_3 contains a term in C_3. If the expression for C_3 given in Eq. (3.4b) is substituted into Eq. (3.4d), it will be found that A_3 is proportional to v_2. Hence it cannot be uncorrelated with v_2.

[9] In fact, any relationship that determined some parameters in terms of others would reduce the number of parameters to be estimated. Our discussion will be limited to this case, which is the simplest, but the generalization poses no special difficulties.

model, any two of the following three consistency conditions,

$$a_1 = a_4, \quad a_2 = a_5, \quad a_3 = a_6, \tag{3.5}$$

will reduce the number of independent parameters to be estimated sufficiently to allow all the remaining parameters to be identified. If all three conditions are imposed simultaneously, the model will be *over*-identified. This permits a partial test of the model.[10]

Only the correlations lagged by a single time unit were used in identifying the regression coefficients in Eqs. (3.4a) and (3.4b). The cross-sectional correlations are used to identify $r_{C_1A_1}$ and the cross-sectional correlations between time two errors $(r_{u_1v_1})$ and between time three errors $(r_{u_2v_2})$. The correlations between time 1 and time 3 scores are not needed to identify the parameters, and because the information they contain is redundant with the information contained in correlations lagged by one time unit, they do not help to overidentify any of the parameters.[11]

Since correlations between time 1 and time 3 scores are not used to identify parameters, it is not essential to the estimation that observations be collected for the same individuals at all three times. If we have two waves of observations for two sets of individuals (e.g., boys and girls) and assume that the same structural equations govern change in both groups, we can impose the same identifying constraints and estimate coefficients provided the corresponding correlations in the two groups are not identical.

To conclude this discussion of the identifiability of the three-wave model, we note that our entire treatment has been premised on the implicit assumption that the system under study has not reached equilibrium. By equilibrium we mean that the correlations between any pair of variables depend only on the lag between them, and not on the wave of observation. In other words, if the system is in equilibrium we will have $r_{A_1A_2} = r_{A_2A_3}$, $r_{A_1C_2} = r_{A_2C_3}$, $r_{A_1C_1} = r_{A_2C_2} = r_{A_3C_3}$, etc.

To see that the analysis in the text depends on the system not being in equilibrium, suppose that $a_2 = a_5$ and $a_3 = a_6$. The discussion of the text leads us to assert that the model is just-identified. The four standardized normal equations are

$$s_{C_2C_1} = a_1 \quad + a_2 s_{A_1C_1} + a_3 s_{A_2C_1},$$

$$s_{C_2A_1} = a_1 s_{C_1A_1} + a_2 \quad + a_3 s_{A_2A_1},$$

[10] With additional waves of data, some of the assumptions about the equality of coefficients can be relaxed without losing identifiability. Thus these assumptions too can be tested.

[11] This is true only because we have assumed that serially correlated errors do not exist. If serial correlations among errors are estimated, lag two correlations can be used for the estimation.

$$s_{C_3C_2} = \qquad a_2 s_{A_2C_2} + a_3 s_{A_3C_2} + a_4,$$

$$s_{C_3A_2} = \qquad a_2 \quad + a_3 s_{A_3A_2} + \overset{\bullet}{a_4} s_{C_2A_2}.$$

It is convenient to express the solutions to these equations in determinant form. If Δ is the determinant of the matrix of coefficients for the right hand members of the normal equations, that is,

$$\Delta = \begin{vmatrix} 1 & s_{A_1C_1} & s_{A_2C_1} & 0 \\ s_{C_1A_1} & 1 & s_{A_2A_1} & 0 \\ 0 & s_{A_2C_2} & s_{A_3C_2} & 1 \\ 0 & 1 & s_{A_3A_2} & s_{A_2C_2} \end{vmatrix}$$

and Δ_j is the determinant obtained by the replacing the jth column of Δ with the column vector

$$\begin{pmatrix} s_{C_2C_1} \\ s_{C_2A_1} \\ s_{C_3C_2} \\ s_{C_3A_2} \end{pmatrix},$$

then $a_j = \Delta_j/\Delta$.

Computation of the determinants shows that each can be expressed as the sum of terms, each of which is a difference between covariances or products of covariances involving the same variables measured in successive waves (e.g., $s_{A_3A_2} - s_{A_2A_1}$, or $s_{C_1A_1}{}^2 s_{A_3A_2} - s_{C_2A_2}{}^2 s_{A_2A_1}$).

There are several consequences of this. First, when the system of equations is in perfect equilibrium, each of these differences will vanish. It follows that the determinants Δ and Δ_j will vanish. Under these circumstances, unique solutions for a_j do not exist, and the equations are underidentified. Intuitively, this is because in equilibrium, covariances between variables measured at times 2 and 3 supply no new information beyond that already contained in the covariances between variables measured at times 1 and 2. Algebraic manipulation of the normal equations shows that a necessary (but not sufficient) condition for the system to be in equilibrium is that $a_1 = a_4$. Thus all parameters must be stationary for equilibrium to prevail.

Second, when the system is not in equilibrium, parameters are identified or overidentified. However, as we show in Chapter 10, when the system is close to equilibrium, measurement error will contribute substantially to differences of terms. As a result, the differences will be unreliable, and standard errors of parameter estimates will be large. These estimates will therefore be imprecise.

To study the behavior of these estimates in the approach to equilibrium, we carried out a series of simulations. We began by assigning arbitrary numerical values to the parameters a_1, a_2, a_3, b_1, b_2, b_3, and $r_{A_1C_1}$, and used the structural equations to compute correlations at later times. This computation was carried out iteratively until the correlations among A_t, A_{t+1}, C_t, and C_{t+1} agreed with the correlations among A_{t+1}, A_{t+2}, C_{t+1}, C_{t+2} to within three decimal places.

Computations of this sort were carried out for 36 different systems of equations. Stability coefficients a_1 and b_1 were allowed to range from .3 to .7; cross-coefficients a_2, b_2, a_3, b_3 were given magnitudes of .1, .2, and .4, and were taken in all combinations of signs, in all possible combinations with the stability coefficients. In every case we assumed arbitrarily that $r_{A_1C_1} = -.9$, far from the equilibrium cross-sectional correlation for these models.

For each model, LISREL IV was used to implement the constancy method for obtaining maximum-likelihood estimates and standard errors of the estimates for each set of three consecutive waves. Thus, for a given set of parameters, the estimation was carried out using the correlation matrix generated by waves 1, 2, and 3, by waves 2, 3, and 4, etc. Standard errors were computed on the assumption that the sample size was 1000.

There was little variation in the properties of the estimates generated by this procedure from one model to the next. The estimates consistently showed that the procedure reproduces the underlying parameters without bias as long as the system is not extremely close to equilibrium; and that the standard errors of the parameter estimates increase, sometimes dramatically, as one moves from the matrix derived from the first three waves to matrices derived from later sets of waves. In a typical model, the first two sets of waves reproduced the model parameters to three decimal places, but the third set of waves yielded discrepancies from the true values as large as 28%, with estimates derived from higher sets of waves even farther off. Standard errors for cross-coefficients that were less than .1 in the first iteration ranged from .4 to .7 in the second, and exceeded 1 in the third. While all parameters were statistically significant at the .05 level in the first iteration, none were in the second.

These results indicate that a set of correlations must be fairly far from equilibrium for the consistency approach to be of practical value. As new organizations form or undergo structural transformations, as individuals enter new environments, as new interventions or constraints on behavior are introduced into groups of individual or collective actors, the consistency approach can be used to estimate causal models. As social systems settle down, the consistency approach becomes less useful.

Given this conclusion, it becomes relevant to the conditions under which systems approach equilibrium, and how rapidly they do so. An analytical investigation of these questions can be found in the technical appendix to this chapter.

A Panel Analysis of Crime Rates and Arrest Rates

To illustrate the approach just outlined, we analyze the relationship between reported annual index crime rates and arrest clearance rates in a sample of 98 U.S. cities for the years 1964 to 1970, on the assumption that the relationships among these variables were not in equilibrium.[12] Although analyses have been carried out separately for each of the index offenses, only the results for total offenses will be reported here.[13] To avoid complicating the exposition, no control variables are introduced in the model we discuss here. We note, though, that our findings about the mutual effects of crime rates and arrest rates on one another were not changed when we controlled for such exogenous influences on crime and arrest rates as the unemployment rate, percent black, and median income of the community.

When panel models involving 1-year lags between observations were examined, it was found that correlations among the crime rates for consecutive years were so highly correlated that individual model parameters could not be determined with precision. Substantively, this means that too little change in the rates occurs over the period of a single year for the impact of a predictor variable on this change to be detected with good accuracy. It follows that if lagged effects of any substantial magnitude exist, they must involve lags of more than 1 year. When models involving lags of 2 or 3 years were studied, this difficulty did not arise. Since the results for 2- and 3-year lag models were similar, only the former will be described here.

The observed correlations among crime rates and clearance rates were fit to a four-wave generalization of the model depicted in Figure 3.3. Standardized crime rates and clearance rates at time t were assumed to depend linearly on one another at times t and t-1, while higher order autoregres-

---

[12] This assumption was justifiable on substantive grounds since the index crime picture began to change radically in the early 1960s. For example, homicide rates began to rise in 1963, reversing a 30-year decline. This trend continued through the mid-1970s, well beyond the time this panel study leaves off. There has been speculation that the change in crime patterns was a consequence of disillusionment with the limited gains won by the civil rights movement, but for the limited purposes of this study the cause of the change is immaterial; what is important is that a change was occurring.

[13] Details regarding the selection of the sample and the definitions of variables are given in Greenberg, Kessler, and Logan (1979).

sive coefficients and cross-coefficients were constrained to be zero. Cross-coefficients linking time 3 with time 4 were constrained to equal those that link time 2 with time 3, but no constraints were imposed on the coefficients that link times 1 and 2. All coefficients representing instantaneous influences were constrained to remain constant. The serial correlations among the error terms were fixed at zero, but the correlations among contemporaneous error terms were left unconstrained and were estimated along with the other parameters. With the constraints just indicated, the model is overidentified with 8 degrees of freedom, and each regression coefficient is just-identified.[14]

Estimation of the model was carried out using the computer program LISREL IV. As outlined in Appendix B to Chapter 2, this program provides maximum-likelihood estimates of regression coefficients in overidentified models, and also has an option that permits the constraints needed to identify the model to be imposed.[15] The fit of a model to the observed data is evaluated by comparing the observed matrix of correlations among the variables with the matrix predicted by the model using the parameter estimates generated by the maximum-likelihood procedure. A chi-square goodness of fit test statistic with degrees of freedom equal to the number of over-identifying restrictions in the model can be used to test the overall adequacy of a given model.

In the case of the four-wave model for total offenses, the value of chi-square is 33.20; with 12 degrees of freedom, this value is statistically significant at the .05 level. Nevertheless, a comparison of the observed and

[14] To find the degrees of freedom, we note that for $n = 8$ variables, there are $n(n - 1)/2 = 28$ independent correlations. We estimate one cross-sectional correlation among the time 1 variables, six stability coefficients, six cross-coefficients, and three cross-sectional correlations, for sixteen parameters in all. The difference between 28 and 16 is 12.

[15] In the model we estimate, no consistency constraints are imposed on coefficients that link time 1 variables. We point out to LISREL users who at some point might want to impose such constraints that it is not immediately apparent when first working with LISREL how coefficients in the beta matrix (which links endogenous variables with one another) can be constrained to equal coefficients in the gamma matrix (which links endogenous variables with exogenous variables), since the signs of all coefficients in the beta matrix are reversed. This can be done, though, by bypassing the gamma matrix entirely and using the "no X" option in the program. This option allows one to treat all observed variables as if they were endogenous. A trivial indeterminacy is resolved by fixing the disturbances of X_1 and Y_1 in the psi matrix to equal their observed values and estimating the covariance of these two variables in the same matrix. This estimated covariance will converge to the observed exogenous covariance once the variances are fixed in this way. All structural coefficients are then specified in the beta matrix. Readers who are not familiar with LISREL will find definitions of the various matrices mentioned here in any of the citations given in Appendix B to Chapter 2.

Table 3.1
Parameter Estimates[a] for a Panel Model of Total Index Offense Rate[b]

	$C \to C$			$A \to A$		
	C_1C_2	C_2C_3	C_3C_4	A_1A_2	A_2A_3	A_3A_4
Parameter						
Parameter estimate	.937*	.924*	.846*	.569*	.675*	.583*

	$A \to C$			$C \to A$				$C \leftrightarrow A$		
	A_1C_2	$A_2C_3 = A_3C_4$	A_tC_t	C_1A_2	$C_2A_3 = C_3A_4$	C_tA_t	C_1A_1	u_1v_1	u_2v_2	u_ev_3
Parameter										
Parameter estimate	−.219	−.228	.295	.926	.849	−1.094	−.346*	−.021	.067	.105

[a] X_iX_j denotes the standardized regression coefficient for the causal effect of variable X at time i on variable Y at time j.
[b] The 95% confidence limits for asterisked parameters do not include zero.

estimated correlation matrices shows the fit to be substantively good. The mean absolute discrepancy of correlations was only .02, with just one difference larger than .10. It is clear from the matrix of differences that a somewhat better fit could be obtained by including small serial correlations among the errors in the model. Since our conclusions were not changed when this was done, we ignore this complication here, and proceed to discuss the parameter estimates we obtained.

Table 3.1 displays the parameter estimates for the stability coefficients of crime rates, the stability coefficients of arrest rates, the effects of arrests on crime, the effects of crime on arrests, and the unanalyzed cross-sectional relationships.

The short-run influence of arrests on crime is measured by the coefficient $A_t C_t$; the lagged effect, by $A_t C_{t+1}$. We note that the estimate of the short-run effect is positive and the lagged effect negative. Both effects are consistent over time, modest in magnitude, and have 95% confidence intervals that include zero. Thus there is no clear evidence in these data that marginal changes in arrest rates affect the crime rate.

The parameters $C_t A_t$ and $C_t A_{t+1}$ provide evidence regarding the effect of crime rates on arrests. Here the instantaneous effect is negative and the lagged effect positive. Although both coefficients are quite large, neither is significant because the standard errors for these estimates are also quite large. We will comment below on the meaning of this circumstance. Again, the lagged and instantaneous effects are separately consistent over time.

The final set of four parameters shows the cross-sectional exogenous correlations among the crime rates and arrest rates at time 1, and the cross-sectional correlations among the error terms for crime rates and clearance rates at times 2, 3, and 4. We note that the correlation between C_1 and A_1 is negative and statistically significant. In a cross-sectional bivariate analysis, this correlation would have been taken as a confirmation of the deterrence thesis, an interpretation which our results call into question.

The correlations among the residual error terms indicate how well the postulated model accounts for the cross-sectional correlations at each of the subsequent time points in the panel (i.e., $r_{C_2 A_2}$, $r_{C_3 A_3}$, and $r_{C_4 A_4}$). Barring measurement error, these cross-sectional correlations would be expected if some exogenous variable that influences both C and A had been omitted from the error terms. Writing the error terms as

$$u_t = d_1 Z + e_1, \quad v_t = d_2 Z + e_2$$

and assuming e_1 and e_2 to be uncorrelated with Z and with one another, we see that the dependence of u and v on Z will result in a nonvanishing correlation between u_t and v_t.

If these correlated error terms proved to be comparable in magnitude to the cross-sectional correlations between C and A, it would indicate that the cross-sectional correlation had been produced primarily by unconsidered exogenous variables. This does not appear to be the case here: the correlated error terms are fairly small and not statistically significant.

Although the parameter estimates of Table 3.1 are consistent with there being no influence of crime on arrests or of arrests on crime, the large standard errors of these estimates mean that effects of substantial magnitude cannot be excluded. These large standard errors, and the high negative correlations between estimates (LISREL provides these correlations as part of its technical output) are clear indications of multicollinearity. The correlations among the crime rates, though not so high as to preclude inversion of the correlation matrix, are nevertheless so high that lagged and instantaneous effects of crime on arrest are not clearly distinguishable.

When multicollinearity is high, the joint effect of lagged and instantaneous variables can be statistically significant even when neither effect is *individually* significant. In other words, the joint confidence ellipse of the collinear variables can exclude zero even when the individual confidence limits all include zero (Hanushek and Jackson, 1977, pp. 132–133). To rule out this possibility, we fixed all cross-coefficients for the influence of crime on arrests, and arrests on crime, at zero and re-estimated the model. As the fit for this model remained good, we were able to accept the null hypthesis. At the .05 significance level there is no evidence that crime affects arrests; or arrests, crime. This finding, together with the very small estimates for the contemporaneous correlations among error terms, suggests that the cross-sectional correlations are due largely to the internal stabilities of the rates themselves and to the initial correlations brought about by historical forces that no longer operate. Omitted exogenous variables and cross-effects among the arrest rates and crime rates make only modest contributions at best.

With this example we see the full power of panel analysis. With four waves of data we were able to use consistency constraints on the parameters to separate out the contributions made to the cross-sectional correlation of the effect of arrests on crime, of crime on arrests, and of omitted exogenous variables. Since we had four waves, it was possible to leave the cross-parameters linking the first and second wave unconstrained, and thereby to check the assumption that the causal effects remain constant over the four waves. These checks showed the assumption of constancy to be quite good. Finally, since the model was overidentified, we were able to carry out a global chi-square test for the model as a whole. A comparison of the observed and estimated correlation matrix indicated the lines along which the model could be improved. We were able to do all of these things *without* having to make the questionable assumptions about the effects of exogenous variables on arrests and crimes that researchers who work with cross-sectional data are forced to make.

Appendix: Approach to Equilibrium in Two-Variable, Two-Wave Panel Models

A convenient starting place for an investigation of the behavior of a panel model over time is Eqs. (3.2a)–(3.2b), which define the model. By eliminating Y_2 from Eq. (3.2a) and X_2 from Eq. (3.2b) we obtain the reduced form equations for X_2 and Y_2. Generalizing to arbitrary times t, it is a matter of straightforward, though at times tedious algebra (omitted here), to compute expressions for the correlations $r_{X_t X_{t-1}}$, $r_{Y_t Y_{t-1}}$, $r_{X_t Y_{t-1}}$, $r_{Y_t X_{t-1}}$ and $r_{X_t Y_t}$.

If p_t represents any of these correlations, the expressions so obtained have the general form

$$p_t = g + hp_{t-1}. \tag{3.A1}$$

Here $g = (b_3 + d_3 + b_1 d_2 + b_2(d_1)/(1 + b_3 d_3)$ and $h = (b_1 d_1 + b_2 d_2)/(1 + b_3 d_3)$.

If $h = 1$, the most general solution to the recursion formula Eq. (3.A1) is $p_t = p_0 + gt$, which increases linearly with t. Since a correlation cannot exceed 1, the structural equation cannot be valid for large time without some change in the values of its structural parameters.

If $h \neq 1$, the most general solution is $p_t = g/(1 - h) + kh^t$, where k is an arbitrary constant to be determined by the initial value p_0 (Goldberg, 1958, pp. 63–67). If the absolute value of h is greater than 1, correlations again will eventually exceed 1, signaling that the system is not approaching equilibrium; if the absolute value of h is less than 1, the system approaches the equilibrium value $g/(1 - h)$ asymptotically.

The speed of approach to equilibrium depends only on h. It follows that approach to equilibrium will be slow when h is close to 1; slow when it is closer to 0. With typical values of structural parameters, the equations will be very close to equilibrium in 3 to 10 time periods (where a period equals the length of the lag in the model).

4

The Components of Change

Thus far we have considered change and its determinants from an explicit causal modeling perspective. At times, though, it is of interest to have simple descriptive statistics to describe change in the sample as a whole, irrespective of how that change was caused. Here one is concerned with such questions as how much change has occurred, whether all members of the sample have experienced change uniformly, and how much of the change in individual positions is governed by their initial position. We will review several different approaches to these problems in this chapter.

The Decomposition of Change

The first issue to be faced in describing change in a population is whether one is substantively concerned only with change at the aggregate level, or whether information about change for individuals is desired. A macroeconomist might want to know, for example, whether spendable income has increased from one year to the next for the population as a whole, without caring whether some people's incomes have changed more than others. On the other hand, someone framing a government policy to deal with poverty might find it relevant to know whether those at the low end of an income distribution tend to remain there for a long time, or whether they tend to be poor only for short periods. To know this, information is needed about income changes for individuals, not just the entire population.

Where information about change in the population as a whole will suffice, the difference of means $\bar{X}_2 - \bar{X}_1$ is an easily interpretable measure of change. Note that it will vanish even if a great deal of change is occurring at the individual level, provided that some cases are decreasing on X enough to compensate for increases in other cases. The difference in standard deviations $s_2 - s_1$ similarly measures the extent to which the dispersion of scores has changed.

If information about change at the individual level is required, we must work with the individual change scores $X_{2i} - X_{1i}$. What is needed is a way of combining the N individual change scores into a single, overall measure of change. We saw in Chapter 2 that the correlation coefficient and regression coefficient between time 1 and time 2 variables could be interpreted to tell us how well change scores were predicted by time 1 scores. But neither coefficient measures the amount of change itself.

The quantity

$$Q^2 = \sum_{i=1}^{N} (X_{2i} - X_{1i})^2/N \tag{4.1}$$

is a convenient measure for the amount of change that has occurred in a population between time 1 and time 2. Since each term enters the sum only after having been squared, the quantity measures the magnitude of change, whether positive or negative. When the static scores are standardized, Q^2 is proportional to $(1 - r_{X_1 X_2})$; when the static scores are unstandardized, Q^2 still declines linearly with $r_{X_1 X_2}$, but the magnitude of Q^2 will reflect the scale of change. If it is desired to measure change in the same units as the static variables, then Q, the root-mean-square of the change scores, may be more useful. But since square roots are inconvenient to manipulate algebraicly, we will carry out our analysis using Q^2.

One question that may arise in studying change concerns the uniformity of change in a population. A change that affects everyone by an equal amount may be quite different sociologically from one whose impact is distributed unequally throughout the population. In the former case, everyone's initial position is preserved by the change, but in the latter case, rankings may be altered. For example, if everyone's income grows by $1000 in a year's time, no one's economic position will have changed relative to others. But if some people's incomes grow while others' decline, this would not be true.

To distinguish these two components of change, we rewrite Q^2 as

$$Q^2 = \sum_{i} \{[(X_{2i} - X_{1i}) - (\bar{X}_2 - \bar{X}_1)] + (\bar{X}_2 - \bar{X}_1)\}^2/N.$$

Making use of the fact that $\bar{X}_2 - \bar{X}_1$ is a constant, and that the quantity in square brackets has a mean of zero, we can write this as

$$Q^2 = \text{var}(X_2 - X_1) + (\Delta \bar{X})^2 = s^2_{\Delta x} + (\Delta \bar{X})^2. \tag{4.2}$$

The second term in Eq. (4.2) represents the contribution to Q^2 of the changing mean. If X changes in such a way that all scores are increased by a constant amount, only this term will contribute to Q^2, and the first term will vanish. On the other hand, if X changes in such a way that its mean remains the same, only the first term contributes. Here there is considerable change at the individual level, but the net effect of this change is to leave the mean of the population unchanged. In this case both terms differ from zero. Their relative magnitude provides information about the character of change: how much is due to change that affects all cases equally, and how much due to change relative to others.

Change in Political Democracy

To illustrate the computations involved in this decomposition, we draw on the work of Bollen (1980) on the comparative measurement of political democracy. Bollen has developed an index of political democracy for use in cross-national research, and has computed the value of this index for a large sample of nations in 1960 and 1965. The sample means for the 112 nations that had no missing values are $\bar{X}_{1960} = 61.48$ and $\bar{X}_{1965} = 56.12$. Were we concerned only with aggregate change we could stop here, noting that between 1960 and 1965 the average level of democracy in these countries had declined only a small amount, roughly 9%.

A consideration of individual change in terms of Q^2, however, shows that a good deal of movement among the countries has taken place even though the average levels of political democracy remained fairly stable over time. To draw this conclusion we need to know that the standard deviations in each year are $s_{1960} = 29.67$ and $s_{1965} = 30.89$, and that the correlation between scores in the two years is .91. Together these descriptive statistics contain all the information needed to compute and decompose Q^2.

The change in mean contributes an amount

$$(\Delta \bar{X})^2 = (61.48 - 56.12)^2 = 31.58$$

to Q^2, while the term

$$\text{var}(X_2 - X_1) = s_1^2 + s_2^2 - 2r_{X_1 X_2} s_1 s_2$$

contributes an amount 166.33. Total Q^2 is the sum of these two contributions, or 197.91. We thus see a fairly substantial amount of change in political democracy among individual nations. But since some nations increased their scores while others declined, the net decline in political democracy in the entire world was not large. Only 15.96% ($= 31.58/197.91$) of the total change arose from an overall decline of political democracy.

Structural Change

Other concerns in the study of change center on the separation of change from lack of change effects in static score measurements (an issue already raised in Chapter 2) and the distinction between "structural" and "individual" change. By structural change, we mean change that can be predicted from initial scores. Change that is not determined by initial scores is "individual" change. Thus, that part of change in income that is predicted by earlier income is structural, while the part that is not is individual.

In breaking down change into these categories, a series of decompositions based on the product-moment correlation between X_1 and X_2 can be instructive.

If we substitute Eq. (2.2) into the formula for the product–moment correlation, we can directly decompose the correlation into parts that are uniquely due to lack of change, and to structured change:

$$r_{X_1X_2} = (s^2_{X_1}/s_{X_1}s_{X_2}) + (s_{X_1\Delta X}/s_{X_1}s_{X_2}). \tag{4.3}$$

Squaring Eq. (4.3), we obtain an expression for X_2 in terms of components that are unique to lack of change, unique to structured change, and an overlap term:

$$r^2_{X_1X_2} = (s^2_{X_1}/s^2_{X_2}) + (s_{X_1\Delta X}/s_{X_1X_2})^2 + 2(s_{X_1\Delta X}/s^2_{X_2}). \tag{4.4}$$

Finally, we can obtain a complete decomposition of the correlation of X_2 with itself simply by adding a component to Eq. (4.3) that is equal to $(1 - r^2_{X_1X_2})$. This term represents that part of the variance of X_2 that has no linear dependence on X_1.

Since the bivariate regression coefficient of X_2 on X_1 is the same as the correlation coefficient when both variables are standardized, these results can be employed directly when working with standardized variables. However, it can be instructive to express Eq. (4.3) in the metric of X_2 rather than as a standardized score. When this is done, $r^2_{X_1X_2}$ is reexpressed as the fraction of the variance of X_2 that is explained by X_1. Denoting the explained variance by $s^2_{\hat{X}_2}$, and the unstandardized regression coefficient of ΔX on X_1 by $b_{\Delta X, X_1}$, we have

$$s_{\hat{X}_2}^2 = r_{X_1 X_2}^2 s_{X_2}^2 = s_{X_1}^2 + (b_{\Delta X, X_1} s_{X_1})^2 + 2b_{\Delta X, X_1} s_{X_1}^2. \qquad (4.5)$$

If we also include the unexplained component of the variance in X_2,

$$s_{X_2}^2 = s_{X_1}^2 + (b_{\Delta X, X_1} s_{X_1})^2 + 2b_{\Delta X, X_1} s_{X_1}^2 + s_e^2. \qquad (4.6)$$

The four components in Eq. (4.6) represent, respectively, the part of X_2 due to lack of change in X over time, the part uniquely due to the structured change in X, the overlap between these two components (due to the fact that individuals with different initial scores will change by different amounts), and the residual change, which is unrelated to X_1.

A careful examination of Eq. (4.6) shows that only the second and fourth components are uniquely due to change. It should not be surprising, then, to find that the change score variance $s_{\Delta X}^2$ is equal to the sum of these two components:

$$s_{\Delta X}^2 = b_{\Delta X, X_1}^2 s_{X_1}^2 + s_e^2. \qquad (4.7)$$

If we bear this equivalence in mind when we interpret a decomposition based on Eq. (4.6), we can in one step study both the components that make up X_2 and those that make up ΔX.

The Structure of Black and White Mobility

To illustrate the algebraic manipulations we have been discussing, we examine intergenerational occupational mobility for white and black males, based on data collected in surveys for the years 1972–1974. Our source for the sample moments in this data set is McClendon (1977).

In our notation, the status of father's occupation will be denoted by X_1 and the status of son's occupation by X_2. The means, standard deviations, and correlations characterizing the distributions for the two groups are shown in Table 4.1. We see from the table that mean occupational status increased more for whites ($\bar{X}_2 - \bar{X}_1 = 11.6$) than for blacks ($\bar{X}_2 - \bar{X}_1 = 6.5$), and that inequality in status grew more for blacks ($s_2/s_1 = 1.62$) than for whites ($s_2/s_1 = 1.06$).

To simplify the notation of Eq. (4.6), we rewrite it as

$$s_2^2 = \underset{\text{(no change)}}{s_1^2} + \underset{\substack{\text{(structural} \\ \text{change)}}}{(b_{\Delta X, X_1} s_1)^2} + \underset{\text{(overlap)}}{2b_{\Delta X, X_1} s_1^2} + \underset{\substack{\text{(nonstructural} \\ \text{change)}}}{s_e^2}.$$

For whites, the "no change" term is $(22.9)^2 = 524.41$. To compute the structural change term, we find from Table 4.1 that $b_{X_2 X_1} = r_{12} s_2/s_1 =$

Table 4.1
Sample Moments for Occupational
Statuses of Fathers and Sons

	Whites	Blacks
$\bar{X}_1$	32.7	19.9
$\bar{X}_2$	44.3	26.4
s_1	22.9	16.3
s_2	24.2	19.8
$r_{X_1X_2}$	.36	.14

Source: McClendon (1977).

$(.38)(24.2/22.9) = .38$, from which $b_{\Delta X, X_1} = .38 - 1 = -.62$. The structural change term is then computed to be $(-.62)^2(22.9)^2 = 201.58$. The other terms are derived similarly. Table 4.2 shows the complete computation for whites and blacks.

Comparing the two sets of figures, we see that while the lack of change component is substantial for both blacks and whites, it is larger for whites (.90 versus .68). This means that the fraction of the variation in sons' occupational statuses determined by their taking up occupations with the same statuses as their father's occupations is larger for white than black sons. Of the change that has occurred, though, most is unrelated to fathers' occupations. This is equally true for whites and blacks (.72 versus .68).[1]

Structural and Exchange Components of Change

We will now consider another type of decomposition that considers structural and individual change somewhat differently. The two ways of conceiving change look deceptively alike, but they are very different. We do not advocate the use of this next decomposition for purposes of causal analysis, although it can be useful in studying the relationship between static score parameters and change. We present it, though, to round out our review of currently available techniques of decomposition.

This alternative approach has been developed by McClendon (1977), who was concerned with the distinction between structural and exchange components of social mobility between generations of fathers and sons.[2]

---

[1] The decomposition of Q^2 derived here is not intended primarily as a tool for the comparison of two populations like blacks and whites, but as a tool for investigating the descriptive properties of change in a single population. A complete understanding of changes in two populations, in fact, would require the comparison of metric autoregression equations and a decomposition of the differences in the mean intrapopulation change scores.

[2] In the social mobility literature the notion of a "structural" component of mobility is

Table 4.2
Decomposition of Variance in Alienation and Change in Alienation

	Whites		Blacks	
	s_2^2	$s_{\Delta x}^2$	s_2^2	$s_{\Delta x}^2$
Lack of change	524.1	—	265.69	—
percentage of total	(.90)		(.68)	
Structural change	201.58	201.58	183.33	183.33
percentage of total	(.47)	(.28)	(.34)	(.32)
Overlap	−650.27	—	−441.05	—
percentage of total	(−1.11)		(−1.13)	
Nonstructural change	509.92	509.92	384.07	384.07
percentage of total	(.87)	(.72)	(.98)	(.68)
Total	585.64	711.50	392.04	567.07
percentage of total	(1.00)	(1.00)	(1.00)	(1.00)

But the approach can be used more generally in any analysis of a changing quantitative variable.

Social mobility researchers have for some time been concerned with distinguishing vertical mobility due to a changing occupational structure from mobility that involves movement within a given structure. It can happen, for example, that over time, the disappearance of farm laborer jobs and the opening up of white collar jobs will cause the occupational status of the population as a whole to rise. This seems like something quite different from mobility in the sense of someone rising or falling relative to others in the population. McClendon's formulation of this distinction is especially appealing.

McClendon notes that absolute mobility, as measured by the differences of unstandardized scores for status, $X_{2i} - X_{1i}$, does not distinguish the two kinds of mobility because an overall rise in statuses in the population can lead to a positive value for the difference even though the ith individual has declined in status relative to others. The difference of standardized scores, $z_{2i} - z_{1i}$, automatically adjusts for changes in the distribution of statuses over time by measuring an individual's position at

more complex than "structural change" as we define it here. It includes, for example, the part of the cross-sectional variance in mobility due to historical forces associated with year of birth and year of entry into the labor force. These complications can be introduced into the decomposition of Eq. (4.6) by estimating a multiple regression equation of X_2 on all these structural influences and using the partial regression coefficients as a set of b coefficients in this equation in place of the single b coefficient in our equation. This extension is trivial, and is not considered in our discussion because it is more an issue of causal modeling than of the description of overall change.

a given time on a scale that is standardized for the mean and variance of the distribution at that time.

By definition, exchange mobility is that part of mobility that does not arise as a result of changes in the overall distribution of statuses. From what we have just said, $z_2 - z_1$ has this property, and is thus a suitable measure of exchange mobility. The time 2 score of someone who has no exchange mobility will be

$$\hat{X}_{2i} = s_2 z_{1i} + \bar{X}_2. \tag{4.8}$$

The distance

$$X_{2i} - \hat{X}_{2i} = s_2(z_{2i} - z_{1i}) \tag{4.9}$$

is thus a measure of exchange mobility in absolute terms. Since *all* the mobility experienced by an individual who has no *exchange* mobility is *structural*, structural mobility can be defined as

$$\hat{X}_2 - X_1 = (s_2 - s_1)z_1 + (\bar{X}_2 - \bar{X}_1). \tag{4.10}$$

Comparison of the Two Approaches

Before illustrating the computation of the different terms in the decomposition of change implied by McClendon's distinction between structural mobility (structural change) and exchange mobility (individual change), we first compare our way of conceptualizing change with McClendon's.

In McClendon's definition, scores at time 1 and time 2 are compared in terms of their values on scales defined by their standard deviations at each time. If X_2 is as many standard deviations away from $\bar{X}_2$ as X_1 is from $\bar{X}_1$, no structural change is said to have occurred. By contrast, in the regression approach, structural influence is defined in terms of the relationship between X_1 and X_2, not in terms of the distributions of the two variables. Given X_{1i}, our best prediction of X_{2i} is

$$\hat{X}_{2i} = b_{X_2 X_1}(X_{1i} - \bar{X}_1) + \bar{X}_2.$$

Thus if the regression coefficient $b_{X_2 X_1}$ is 1.5, the distance of each score from the mean at time 2 is predicted to be 1.5 times the distance of the score from the mean at time 1.

It can be informative to rewrite the expression for $\bar{X}_2$ as

$$\hat{X}_{2i} = b_{X_2 X_1} s_1 z_{1i} + \bar{X}_2 \tag{4.11a}$$

$$= s_{\hat{X}_2} z_{1i} + \bar{X}_2. \tag{4.11b}$$

But since

$$s_{\hat{X}_2} = r_{\hat{X}_2 X_2} s_2 = r_{X_1 X_2} s_2 ,$$

we can express

$$\hat{X}_{2i} = r_{X_1 X_2} s_2 z_{1i} + \bar{X}_2 . \tag{4.11c}$$

A comparison of Eqs. (4.8) and (4.11c) shows that the first terms differ by a small but important factor, $r_{X_1 X_2}$. This difference can be traced to the different assumptions of the two approaches. Both begin with a baseline assumption that every individual's time 2 score will be $\bar{X}_2$, and then utilize the information that at time 1 the individual's score was at the z_{1i}th standard deviation from the time 1 mean to improve on this prediction. Since it is possible to calculate the relationship between time 1 and time 2 scores, this additional contribution is given by the initial z score multiplied by the correlation coefficient $r_{X_1 X_2}$. The factor s_2 expresses this contribution in the metric of time 2 scores. In the McClendon approach, on the other hand, the assumption is that the time 2 scores will be at the very same standard deviation level as the time 1 score.

The difference between the two formulas is far from inconsequential numerically. Suppose, for example, that at time 1 an individual is one standard deviation above the mean ($z_{1i} = 1$), and that the correlation between X_1 and X_2 is .5. Then the regression approach leads us to expect the individual's score at time 2 to be $(.50)s_2(1) + \bar{X}_2$, while the Mc-Clendon approach leads us to expect $s_2 + \bar{X}_2$. If the time 1 score is 3 standard deviations below the mean and the correlation between X_1 and X_2 is .2, the time 2 score is predicted to be $(.2)(3) = 0.6$ standard deviations below the time 2 mean in the regression-based approach, but 3 standard deviations below the mean in the McClendon approach. As we see from these examples, the two estimates will be quite different when the stability of scores is low.

The choice of one of these approaches over the other is largely a matter of conceptualization. For causal analyses we favor the regression-based approach because it is more consonant with our way of thinking about causal processes. In McClendon's approach, rather awkward causal assertions must be made. Where we assert that X_1 has a certain causal influence on X_2, or on change in X, McClendon must assert that the changing dispersion of X exerts an influence on the change in X. But it is difficult to think of how the changing disperson of a variable can have a causal influence on that variable.

Despite the greater usefulness of the regression approach when interest focuses on the influence of initial scores on change, McClendon's approach can be used to good effect when one is concerned with the contribution to overall change of changes in the individual parameters that characterize the distribution (mean, variance). To see this, we express Eq. (4.2) in terms of the static scores:

$$Q^2 = \sum_i (\Delta X_i)^2 / N = (\Delta \bar{X})^2 + (s_1^2 + s_2^2 - 2r_{X_1 X_2} s_1 s_2). \qquad (4.12)$$

First consider the case in which the standard deviation of the distribution is constant over time ($s_1 = s_2$) and the standardized scores z_i do not change (i.e., relative positions in the distribution are maintained). This implies that $r_{X_1 X_2} = 1$. In this special case, Eq. (4.12) reduces to

$$Q_{\Delta X}^2 = \sum_i (\Delta \bar{X})^2. \qquad (4.13)$$

Next consider the case in which the mean of the distribution and the standardized scores remain constant, but the standard deviations may vary with time. Equation (4.12) here takes the form

$$Q_{\Delta s}^2 = (s_2 - s_1)^2. \qquad (4.14)$$

McClendon's measure of structural change is the sum of these two terms:

$$Q_{\text{structural}}^2 = (\Delta \bar{X})^2 + (s_2 - s_1)^2. \qquad (4.15)$$

It receives contributions from two sources: the changing mean of X and the changing standard deviation of X.

Now consider the case when $\bar{X}$ and s remain constant, but the relative positions of individuals may change between time 1 and time 2 ($r_{X_1 X_2} < 1$). In this case Eq. (4.12) reduces to

$$Q_r^2 = 2s_2^2(1 - r_{X_1 X_2}). \qquad (4.16)$$

This is McClendon's measure of exchange, or individual, change.

Since the dispersion of a distribution can change at the same time relative positions are changing, the complete decomposition of Eq. (4.12) contains an overlap, or interaction term:

$$Q^2 = Q_{\Delta X}^2 + Q_{\Delta s}^2 + Q_r^2 + Q_{\text{overlap}}^2. \qquad (4.17)$$

It is a trivial but tedious exercise in algebra to show that

$$Q_{\text{overlap}}^2 = 2s_2(s_1 - s_2)(1 - r_{X_1 X_2}). \qquad (4.18)$$

This expression makes clear that Q_{overlap}^2 differs from zero only when the standard deviation and relative positions of scores are simultaneously changing between time 1 and time 2. Note that when the standard deviation is increasing with time, this term will be negative. What does this mean? Apart from a factor of 2, Q_{overlap}^2 is equal to the covariance of $\hat{X}_2 -$

X_2 and $\hat{X}_2 - X_1$ (the computation can be carried out using Eqs. [4.9] and [4.10]). But $\hat{X}_2 - X_1$ represents an individual's exchange mobility, and $\hat{X}_2 - X_1$, an individual's structural mobility. Thus when the dispersion of the distribution is increasing, these two types of inequality are negatively correlated. Since persons with high initial scores are favored by high structural mobility, but disfavored by high exchange mobility, an increasing dispersion tends to reduce the negative impact of exchange mobility on those who have high scores initially.

The decomposition given in Eq. (4.17) shows what part of the overall change in X is due to the changing mean, the changing standard deviation, the changing relative position of scores in the distribution, and the simultaneous change in standard deviation and relative position of scores.

Applications

McClendon devised his decomposition as an aid to the study of occupational mobility, and illustrated his approach by analyzing the figures given in our Table 4.3.[3] Using Eqs. (4.13)–(4.18), we can estimate the different contributions to Q^2. The results of these computations are displayed in Table 4.3.[4] Exchange mobility is quite high for both groups, considerably larger than structural or overlap mobility. The change in average status is larger for whites than for blacks, while the dispersion in statuses increases more for blacks than for whites. As the first effect outweighs the second numerically, structural change is larger for whites than for blacks. On the other hand, overlap change is larger for blacks

Table 4.3
McClendon's Decomposition of Mobility[a]

Components of Change	Whites	Blacks
$Q_r^2 = Q_{exchange}^2$	726.19	674.31
$Q_{\Delta\bar{x}}^2$	134.56	42.25
Q_s^2	1.69	12.25
$Q_{structural}^2$	136.25	54.50
$Q_{overlap}^2$	−40.27	−119.20
Q_{total}^2	822.17	609.61

[a] The data from which these figures have been computed were taken from McClendon (1977, p. 66).

[3] In addition, McClendon analyzed mobility for males interviewed in 1962 and for females interviewed in the years 1972–1974.

[4] The entries in McClendon's table (1977, p. 66) are square roots of Q^2, and thus cannot be compared directly with the entries in our table.

than for whites, reflecting the greater increase in dispersion of scores for blacks. In both cases, the overlap term contributes negatively, reducing the effect of structural change in giving an advantage to those whose fathers have high status occupations.

Although McClendon's approach was developed to analyze social mobility, it can be used to good advantage whenever the researcher is concerned primarily with changes in the parameters describing the cross-sectional distributions. For example, Collver and Semyonov (1979) used it to study the changing characteristics of suburban areas. Special theoretical interest centered on the following questions:

(1) Have the averages of SES characteristics of suburbs increased, decreased or stayed the same. . . ? (2) Have the differences between suburbs widened or narrowed. . . ? (3) To what degree have the individual suburbs maintained or changed their relative positions in the status hierarchy. . . ?

McClendon's decomposition approach is ideally suited to questions such as these and was used by Collver and Semyonov to explore these questions of suburban persistance and change.

When the researcher is concerned primarily with the causal structure of changes in a variable between two times, rather than with changes in the cross-sectional distributions, we recommend using the regression-based decompositions developed here.[5] Unlike McClendon's decomposition, the regression-based decompositions are interpretable in terms of causal links between X_1 and X_2.

---

[5] The reader might find it instructive to work out both decompositions for the data on SES and alienation given in Chapter 2, and the data on change in political democracy given in this chapter. All the information that is needed for the computations is given in both cases.

5

Cross-Lagged Panel
Correlations

Cross-lagged panel correlations (CLPC) is a technique developed some time ago by Campbell (1963) to assess the relative importance of the variables X and Y in a causal analysis by comparing the lagged zero-order correlations $r_{X_1Y_2}$ and $r_{Y_1X_2}$. Though subjected to withering criticism (Duncan, 1969a, 1972; Heise, 1970; Rogosa, 1979, 1980), the technique has been widely used in psychology and occasionally in sociology (a score of applications are cited in Sims and Wilkerson, 1977 and Rogosa, 1980; and the bibliographies in these sources are by no means complete). More recently, the technique has been proposed as a test for spuriousness. We examine both applications of the method.

Background

Campbell (1963) based his original formulation of the cross-lagged panel technique on Lazarsfield's (1948) method for analyzing change in the 16-fold table.[1] In the latter method, cell frequencies in the 4×4 table for cell frequencies of two dichotomized variables measured at two points in time are used to draw inferences about the direction of causal influences between these two variables. Campbell extended this logic to interval-level variables by proposing that the cross-lagged correlations $r_{X_1Y_2}$ and $r_{Y_1X_2}$ be compared. Rather than study the various kinds of complex causal

---

[1] A recent discussion of Lazarsfeld's approach in the context of the Goodman log-linear model can be found in Kessler (1977a).

processes considered by Lazarsfeld (the generation and preservation of effects among them), Campbell considered only a single question: is X more important than Y or Y more important than X in bringing about an observed cross-sectional association between X and Y?

Before we consider Campbell's logic in detail, we note that as a means of disentangling causal priorities in panel data, CLPC is distinct from the approach we have outlined in Chapter 3 in that it focuses on the *relative* magnitude of the processes "X causes Y" and "Y causes X." It is also distinct in that *individual parameters for these distinct processes are not estimated*. Only the difference between the cross-lagged correlations is considered.

CLPC as a Tool for Causal Analysis

As initially formulated by Campbell (1963) and Campbell and Stanley (1963), the CLPC technique was based on the following line of reasoning: if X is a stronger cause of Y than Y is of X, the correlation $r_{X_1Y_2}$ should be larger in magnitude than the correlation $r_{Y_1X_2}$. As a test of causal influences, this simple comparison has several problems that soon became obvious to Campbell and his students.

Temporal Misspecification

One class of these problems involves temporal misspecification, described more narrowly by Kenny (1973) and Cook and Campbell (1979) as "temporal erosion." The difficulty here is that misspecified causal lags can create significant observed cross-lagged correlation differences even when they do not exist in the true causal processes. Kenny's example asks us to consider the possibility that X has been measured at times .5 and 1.5, while Y has been measured at times 1.0 and 2.0—as could happen through aggregation.[2] Then the cross-correlation of the earlier measurement of X with the later measurement of Y will involve a lag of two time units, while the cross-correlation of earlier Y with later X will involve an interval of .5 time units. If it is true that, all else being equal, cross-correlations attenuate with increasing time intervals, we will obviously bias our analysis in this instance in favor of the hypothesis that Y causes X.

[2] For instance, school grade point average at time 1 is actually a measure of performance over the last several months before the grade is recorded, while self-reported happiness with school is an attitude that has a strong "today" bias. A CLPC analysis of these two variables would probably exhibit a temporal erosion bias.

As it happens, correlations do not always erode monotonically with time (see Chapter 8), and therefore temporal erosion need not *always* be a problem. But it is common enough to be a source of serious concern. Therefore it is important to consider carefully the time interval over which one expects each of the variables to influence the other. A misspecification of this interval (as would occur if the appropriate causal lags are of different lengths and this has not been taken into account in data collection and model specification) can seriously bias the outcome of the cross-lagged comparison.

Nonstationarity

Another class of problems noted in Campbell's later writings (Cook and Campbell, 1979) concerns the consistency of effects across time, something that Kenny (1973, 1975, 1979) calls "stationarity." If it happens that the reliability of one of the variables changes through time, or if the relationship between the measured variable and an underlying construct to which we are making some inference changes through time, the cross-lagged correlations will attenuate in ways unrelated to the relative causal influences.

This process, too, can bias the CLPC technique.

Models of Lagged Causal Influence

Bohrnstedt (1969), Duncan (1969a, 1972, 1975), and Heise (1970), all working within a path analysis framework, showed that CLPC is inappropriate as a test of relative causal influence because lagged cross-correlations are determined not only by differences in the causal influences from X to Y and from Y to X, but also by the differential stabilities of X and Y. This problem is by now so well known that even the strongest advocates of CLPC accept the criticism as valid. Yet, since novices continue to use CLPC blissfully unaware of this criticism, it will be instructive to review the difficulty briefly here.

Consider Figure 5.1, in which a path diagram for a cross-lagged panel model is illustrated. Let us assume that the true causal connections between X and Y are as shown. Each of these variables is causally influenced by the earlier scores with path coefficients a, b, c, and d, as in the figure. From the fundamental theorem of path analysis, the cross-lagged correlations can be expressed as follows:

$$r_{X_1Y_2} = b + dr_{X_1Y_1}, \tag{5.1a}$$

$$r_{Y_1X_2} = c + ar_{X_1Y_1}. \tag{5.1b}$$

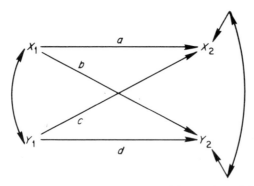

Figure 5.1. A two-wave, two-variable panel model.

We see from these expressions that the cross-lagged correlations receive contributions from the true causal coefficients b and c; but, in addition, they receive contributions proportional to the stabilities a and d. Only when the stabilities are equal does equality of the causal coefficients b and c follow from equality of the cross-correlations. Usually, though, the stabilities of the two variables will be unequal. In that case, no inference about the relative strength of the two effects is possible. Indeed, not even the signs of the true causal influences can be detected here.

Recognizing this, Pelz and Andrews (1964) suggested that differential stability be controlled by means of partial correlations. Yet this approach is simply a less efficient way of doing the causal modeling of partial regression equations that we described in earlier chapters. A partial regression approach is to be preferred to a partial correlation approach because the regression coefficients are parameters in explicit structural models, while partial correlation coefficients are not. Furthermore, a regression approach allows estimation of a far wider variety of models than the simple cross-lagged model suggested by Campbell (Duncan, 1969a, 1972, 1975).

As a result of these considerations it is now generally agreed that *if the purpose of one's analysis is to estimate causal parameters*, then a multiple regression approach is to be preferred to a CLPC approach. If one wants to do so, the cross-lagged partial regression coefficients can be compared in a variety of ways. Or, preferably, the correlation $r_{X_2Y_2}$ can be decomposed into its component causal parts so that the relative influences of the cross-coefficients can be studied (Alwin and Hauser, 1975).

As we stated above, even the staunchest advocates of CLPC recognize this. Yet at times, this recognition is seemingly forgotten in the course of analysis. Consider, for example, the correlation matrix of Table 5.1. These correlations, found in the work of Kidder *et al.* (1974) on rates of

Table 5.1

Panel Correlations for the Relationship between Burglary
Rates and Police Force Strength for a Sample of
U.S. Cities[a]

	B_1	B_2	P_1	P_2
B_1		.89	.47	.43
B_2			.35	.39
P_1				.86
P_2				

Source: Kidder *et al.* (1974).
[a] B represents burglary rate; P represents the size of police force.

burglary and numbers of police officers in a sample of American cities, were analyzed by Kenny (1979, p. 236), who uses the CLPC technique to address the question of which is causally more important: the presence of the police in reducing crime, or the effect of crime on police force build-up.

Commenting on these correlations, Kenny states:

> At first glance it appears that burglaries cause an increase in the number of police. An alternative "law and order" explanation is that the number of police causes a decrease in the number of burglaries. Both hypotheses are equally plausible. The data are not consistent with two other hypotheses: police increase burglaries or burglaries decrease the number of police. These later hypotheses are not ruled out but their effects, if they exist, are *overwhelmed* by the effects of one or both of the former hypotheses [Kenny, 1979, pp. 235–236; emphasis added].

Actually, if we take the high stabilities of police and burglaries across the twelve months of the panel into account by computing the *path coefficients* for the effect of B_1 on P_2 and the effect of P_1 on B_2, we find that they are not meaningfully different from zero.[3] There is no persuasive evidence, then, for *any* of the hypotheses Kenny mentions, and certainly none that any of the causal effects is "overwhelming."

In this example, the use of CLPC led a seasoned researcher to conclude erroneously that pronounced causal effects were present in the data when they were not. The opposite fallacy, of failing to conclude that genuine causal effects are present, is equally possible. Consider, for example, the correlations displayed in Table 5.2. These correlations show the relationships between IQ scores and scores on a composite index of scholastic achievement for 5495 students tested in fourth grade, and again in sixth grade.

---

[3] The standardized path coefficient for the effect of B_1 on P_2 is found to be .03, while the corresponding coefficient for the effect of P_1 on B_2 is -0.09.

Table 5.2
Panel Correlations for IQ and Scholastic Achievement (ACH) in
Grades 4 and 6 ($N = 5495$)[a]

	IQ_4	IQ_6	ACH_4	ACH_6
IQ_4		.83	.78	.75
IQ_6			.73	.77
ACH_4				.80
ACH_6				

Source: Crano, Kenny, and Campbell (1972).
[a] The correlations in the table have been rounded off to two
decimals.

Working on the assumption that the model in Figure 5.1 is valid for this
example (that IQ and scholastic achievement are mutually dependent) the
standard causal modeling approach proceeds by estimating the causal
influences via the methods described in Chapter 3. In the present case,
assuming that the causal influences are lagged, the standardized regression coefficients are estimated to be

$$IQ_6 = .52IQ_4 + .21ACH_4,$$
$$ACH_6 = .55ACH_4 + .31IQ_4.$$

The stabilities of the two variables are moderately high; and mutual
causal influences between the two variables are both positive and moderate in magnitude, with the effect of IQ on achievement somewhat larger
than the effect of achievement on IQ. Although this set of findings is susceptible of more than one interpretation, perhaps the simplest is that
achievement is a measure of acquired skills, and these influence later IQ
scores, even when earlier IQ scores are taken into account. This finding is
of no small interest.

Yet the CLPC test, when applied to these same data, could well lead us
to overlook this finding. A comparison of the cross-lagged correlations

$$r_{IQ_6ACH_4} = .7273 \quad \text{and} \quad r_{ACH_6IQ_4} = .7467$$

yields the difference .0194. As it happens, these data were analyzed by
Crano, Kenny, and Campbell (1972) using just this approach. Although
the extremely large sample size allowed these investigators to conclude
that the cross-lagged difference is significant and thus that intelligence
causes achievement more than achievement causes intelligence, with a
sample of the size social scientists typically encounter (a few hundred and
seldom more than a thousand) the difference would not be significant.
Under this circumstance, the CLPC test is mute, and no conclusion can
be drawn about causal inferences. Even when the correct inference about

relative causal importance is drawn, the more detailed information about the specific magnitudes of the individual causal effects, routinely available when the regression approach is used, is overlooked.

These examples make clear that an analysis of the data from a path analysis perspective (or, more generally, in terms of an explicit structural equation model) is superior to CLPC in its ability to estimate causal effects and test causal hypotheses. For this reason, methodologists now agree that CLPC should *not be used* to model causal processes among panel data when the analyst is concerned only with relations among the observed variables.

We emphasize this point, for we have all too often heard it said by researchers working with CLPC that it really makes little difference if one uses CLPC rather than a structural equation approach to model causal processes. As the example just worked out illustrates, this view is entirely incorrect. The assessment of cross-lagged correlations tells us absolutely nothing about the magnitudes or the signs of cross-lagged causal coefficients. When outside considerations are brought to bear, as in the work of Rozelle and Campbell (1969), who tried to establish a "no-cause comparison base," we merely have an inferior attempt to approximate the control for stability afforded directly via structural equation modeling.

CLPC as a Test for Spuriousness

Apart from the fallacious application of CLPC to assess the relative importance of the two processes $X \to Y$ and $Y \to X$, another, currently more popular rationale for using CLPC techniques instead of a regression analysis has been proposed: to test the null hypothesis that the observed relationships among X and Y are entirely due to some common, unmeasured cause of both observed variables. In *cross-sectional* analyses, it is generally not possible to draw any conclusions whatsoever about spurious sources of correlation. But certain limited inferences about spuriousness can be drawn from panel data. For some of these inferences, the comparison of zero-order cross-lagged panel correlations provides a critical test of the null hypothesis.

This remarkable feature of CLPC was first noted by Duncan (1972) and Kenny (1973), who, in the course of their investigations of cross-lagged panel correlation differences in a variety of underlying causal models, noted that under certain restricted conditions, the correlations $r_{X_1Y_2}$ and $r_{Y_1X_2}$ are expected to be equal.

As one example, suppose that two observed variables X and Y are related only because another unmeasured variable Z is a common cause of

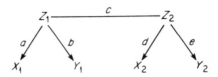

Figure 5.2. A two-wave, two-variable model of complete spuriousness.

both of them. A situation of this sort is depicted in Figure 5.2. If the relationship between Z and the observed variables is stationary, we will have the equalities $a = d$ and $b = c$ for the coefficients in the figure. It follows that the cross-correlations $r_{X_1Y_2}$ and $r_{Y_1X_2}$ will be equal, since the normal equations for these correlations are

$$r_{X_1Y_2} = ace \quad \text{and} \quad r_{Y_1X_2} = bcd.$$

If $a = d$ and $b = e$, the expressions for the two correlations will be equal, *quod erat demonstratum*.

This conclusion remains valid under conditions of proportional change in the causal relations between Z and the observed variables (that is, when the ratios a/b and d/e are equal), and in the presence of multiple Zs, all of which are stationary or proportionately stationary. Moreover, the equality is still valid under conditions of measurement error in X or Y (or both) as long as the reliability of each variable is constant over time. These results are discussed in further detail in the Appendix to this chapter.

The appealing feature of this approach is that it permits one to draw limited conclusions about uncontrolled sources of correlation, something that is usually possible only by means of randomization in experiments. For this reason, the CLPC approach has recently enjoyed something of a resurgence of popular interest.[4]

The use of a CLPC comparison for purposes of evaluating a null hypothesis involving total spuriousness is quite legitimate, though two caveats regarding its use for this purpose must be entered. First, the CLPC test cannot be used to confirm complete spuriousness when causal influences are instantaneous rather than lagged. This situation is depicted in

---

[4] This is especially true in psychology, a surprising fact given that an extremely lucid critique of CLPC appeared in one of the major psychological journals some years ago (Duncan, 1969a). It seems that the special interest among psychologists stems from their desire to approximate experimental controls in nonexperimental work. Presumably this is so because experimental research is much more common in psychology than in the other social sciences.

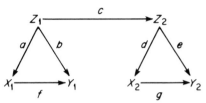

Figure 5.3. A two-wave, two-variable model of partial spuriousness.

Figure 5.3. Since, in the notation of the figure,[5]

$$r_{X_1Y_2} = ac(e + dg) \quad \text{and} \quad r_{Y_1X_2} = cd(b + af),$$

we will have $r_{X_1Y_2} = r_{Y_1X_2}$ provided that the stationarity conditions $a = d$, $b = e$, and $f = g$ are valid. This is so even though X influences Y. Thus the CLPC test is *totally* unable to detect violations of the null hypothesis in this circumstance. The test works only if causal influences are lagged by an amount that is approximately equal to the length of time between the two waves of observations. When the causal influences are approximately simultaneous, the test fails.

Second, a consideration of the cross-lagged panel correlations *alone* is, in some instances, too narrow a test. By beginning with an explicit causal model, such as that diagrammed in Figure 5.2 or 5.3, it may be possible to derive several different tests of the null hypothesis. For example, under conditions of complete spuriousness, it is quite easy to show for the model of Figure 5.2 that the equality $r_{X_1X_2}r_{Y_1Y_2} = r_{X_1Y_2}r_{Y_1X_2}$ holds, but is in general not valid when X and Y are themselves causally related.[6] This, too, can serve as a test for complete spuriousness. Indeed, it is a more powerful test of spuriousness because *its validity does not depend on the stationarity assumption.*

The value of this second test can be seen by returning to the correlations in Table 5.1. Here the CLPC test involves the comparison of $r_{B_1P_2} = .43$ with $r_{P_1B_2} = .35$. Since the difference between these correlations, .08, is fairly small, one might be hesitant to reject the null hypothesis of complete spuriousness, especially if the sample size is not large. (The statistical power of the significance test for this difference is low.) The second

---

[5] For simplicity, the model in Figure 5.3 assumes that X influences Y, but Y does not influence X. None of our conclusions change if the relationship between X and Y is assumed to be reciprocal.

[6] This equality also holds for the model of Figure 5.3 even when X and Y are causally related, thus our earlier warnings about the dangers of causal misspecification are relevant for this test as well. For the model of Figure 5.3 there is no test of complete spuriousness. *C'est la vie!*

test involves the comparison of

$$r_{B_1B_2}r_{P_1P_2} = (.89)(.86) = .7654$$

with

$$r_{B_1P_2}r_{P_1B_2} = (.43)(.35) = .1505.$$

As the difference between these figures is quite substantial, one may reject the null hypothesis with considerably greater confidence.

This example makes clear that the CLPC test is certainly not the only, nor even the most useful, test of complete spuriousness in a given set of data. To focus narrowly on this one test, and to reify it as a method unto itself, is to be far too narrow in one's perspective on causal modeling. A more explicit reference to the presumed underlying causal structure and a derivation of model-testing predictions from this structure should be preferred by anyone who wishes to test for spuriousness in panel data.

CLPC as a Control for Spuriousness

As we noted earlier, CLPC is not, and was never intended to be, a method for estimating the structural parameters of an explicit causal model. It is a method for testing a null hypothesis. As a result, one should not think of CLPC as a way to *control* for spuriousness in a causal model involving X and Y. By control, we refer to the statistical operation that adjusts results in a causal model for the exogenous influences of theoretically confounding variables. CLPC cannot be used as a means of control in this sense. It either supports the hypothesis that the observed relationships among X and Y are *completely* due to some set of unmeasured Z, or else rejects the hypothesis of complete spuriousness.

This point has sometimes been misunderstood by researchers who have mistakenly regarded the spuriousness test as a substitute for the control for unmeasured variables that randomization provides in experimental research. Thus one occasionally sees researchers interpreting cross-lagged correlation differences as "net" measures of relative causal influence, purged of the spurious influences of unmeasured variables (Humphreys and Stubbs, 1977). The reasoning here is that cross-lagged correlations will be equal under a condition of pure spuriousness; hence any difference between these two correlations reflects the differential causal influences of X on Y and Y on X. Regrettably, some reflection on the underlying structure of the model shows that this reasoning is completely unfounded.

To see this, consider Figure 5.4, where we have diagrammed a situation

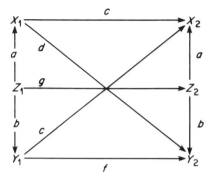

Figure 5.4. A two-wave, two-variable, model of partial spuriousness. A realistic model would also include an exogenous correlation between X_1 and Y_1, but this term can be absorbed in the product (ab) for the purposes of this illustration.

of partial spuriousness. Expressing the observed correlations in terms of the underlying parameters, we have

$$r_{X_1Y_2} = d + abg + fr_{X_1Y_1}, \tag{5.2a}$$

$$r_{Y_1X_2} = c + abg + er_{X_1Y_1}. \tag{5.2b}$$

The first term in the right-hand member of each equation is the direct causal influence of one observed variable on the other. The second term represents the influence of the spurious variable Z. The third term expresses the influence of stability in the time 2 variable in combination with the exogenous correlation between X_1 and Y_1.

Note that (1) the second terms, representing the spurious components of the two correlations, are identical, a result consistent with what we already know about the case of complete spuriousness, and (2) in the absence of differential stability the cross-lagged difference will reduce to the difference between the cross-causal effects c and d.

Once these two points are grasped, it should be obvious that the cross-lagged difference will generally not be equivalent to the true causal difference. The difficulty is exactly the same as the one we observed earlier in our discussion of CLPC as a tool of causal modeling: the stability contributions [the third terms in Eqs. (5.2a) and (5.2b)] confound the true causal influences of X and Y on one another.

In our earlier discussion, we found that this problem could be handled in a structural equation approach. By controlling for time 1 values of X and Y, the partial regression coefficients for the lagged influence of X on Y and Y on X could be isolated. This procedure, then, adjusted for possible differences in the stabilities of X and Y. In the present case, the comparable procedure would require controlling X_1, Y_1, and Z. In terms of Eqs. (5.2a) and (5.2b), the stability contributions $fr_{X_1Y_1}$ and $gr_{X_1Y_1}$ cannot be isolated and adjusted as they could in the previous example unless the product abg is known. This product, however, cannot be estimated from observations on X and Y alone.

To illustrate the lack of correspondence between cross-lagged causal coefficients and cross-lagged correlations in the model of Figure 5.4, we note, referring to Eqs. (5.2a) and (5.2b), that (1) when the stability coefficients e and f are sufficiently large, the cross-correlations can be positive even when the cross-lagged causal coefficients are negative; (2) when the cross-causal coefficients are equal we will still find that the cross-correlations will be unequal as long as the stability coefficients are unequal (as they usually are); and (3) when one cross-causal influence is larger than the other (say, d is larger than c), the relative magnitudes of the cross-correlations can be reversed if the stability of the causally dominant variable is the stronger of the two stability coefficients (in this illustration, when e is larger than f).

The moral of this lesson has been made explicit by Kenny (1973), in a discussion of the limitations to the CLPC technique: "If the researcher suspects that certain unobservables affect both variables in the model, he should include such variables in the model." This is necessary because the CLPC technique *cannot* be used to control for them.

It is a measure of the confusion surrounding this simple point that several investigators have recently begun their analysis of two-wave, two-variable panel data by considering the cross-lagged correlation difference test. After rejecting the hypothesis of *complete* spuriousness, they have then gone on to interpret lagged correlations between X and Y as if no spuriousness at all were present (e.g., Kahle and Berman, 1979). The reasoning employed here seems to be something along the following lines: "Let's first make sure that the *whole thing* isn't spurious. Once we've settled that issue in the negative, let us go ahead and estimate the parameters among the observed scores on the assumption that *no part* of the observed correlations is spurious."

But to readers who have followed our argument so far, it should be obvious that situations of *complete* and *partial* spuriousness are by no means the same, and that the rejection of the complete spuriousness hypothesis tells us nothing at all about the possibility of partial spuriousness. To show through the CLPC test that complete spuriousness does not account for the observed relationships among the panel variables says nothing whatsoever about the possibility that, say, 50% of each observed relationship is spurious.

The Comparison of Contending Models

In a recent review article, Kenny and Harackiewicz (1979, p. 377) claim that "The major purpose of cross-lagged analysis is not so much to test the null hypothesis of spuriousness . . . but rather to demonstrate causal predominance." They go on to note that Duncan and others have shown, on the basis of models similar to the one shown in Figure 5.1, that the relative size of cross-lagged correlation differences does not necessarily parallel the relative size of cross-lagged causal differences. They note, quite correctly, that this result is true only within the context of a particular causal model and that Kenny (1973) has proposed a different type of underlying causal model, in which only one-way causal influences appear (if X causes Y then Y does not also cause X). In this model, the cross-correlation difference will have the same sign as the difference between causal coefficients.

Mindful of this, Kenny and Harackiewicz maintain that "The test of the relative merits of each [contending model] will be based not on statistical analysis but on the degree to which each leads to interpretable and scientifically meaningful results." We would add to this the observation that the degree to which results are scientifically meaningful hinges on the plausibility of the assumptions in each model. When one considers the model described by Kenny (1973), it becomes apparent that it requires that X_1 and Y_1 are causally related only through the common influence of some unmeasured variable Z. If this is so, then one can evaluate the causal influence of X on Y by using the CLPC test. But consider this: if X_1 does in fact cause Y_2 (or, for that matter, if Y_1 causes X_2) there will be some residual relationship between X_1 and Y_1 after controlling for Z, since presumably this cross-effect will have persisted through time. That is, X_0 has caused Y_1 and has also caused X_1, thus making X_1 and Y_1 exogenously related over and above the influence of Z. Thus, in order to test the possibility that X_1 causes Y_2 in his model, Kenny implicitly assumes that X_0 has not caused Y_1! We advance this as an illustration of an implausible model.

An easy way to guard against the use of an implausible model such as this is to begin with a substantive model of the causal processes presumed to be operative in one's data. When there is doubt as to whether a given influence should be included in the model, consider it as if it were present. This applies to the case of unmeasured causes of spurious association among the observed variables as well.

Once one has a model such as this, it is a relatively simple matter to decide if any individual parameters can be identified, and, if so, the sorts of assumptions about the presence, absence, or consistency of causal processes that must be made to achieve identifiability. At times, of course, it will become clear that no individual parameters can be identified without making assumptions that the analyst finds too implausible to accept. For instance, in a two-wave panel analysis, one might believe, even after controlling for all the sources of spurious correlation contained explicitly in the data set, that important unmeasured common causes of X and Y remain unconsidered. When this is so, it will generally not be possible to identify any parameters in a model of this sort in the absense of extremely restrictive assumptions about the parameter values relating the unmeasured common causes to the observed scores—TANSTAAFL.[7]

When seen for what it is, a very limited test of a narrowly specified null hypothesis without any serious implications for parameter estimation in structural models, it seems unlikely to us that researchers will have a great deal of use for CLPC. It may possibly be of some use to an analyst who is working with previously collected data and who wants to distinguish the possibility of complete spuriousness from a model in which at least some causal processes between X and Y are at work, even though rejection of complete spuriousness will permit nothing whatever to be said about the separate or relative importance of these causal processes. It is inconceivable to us, though, that a panel analysis would be specifically designed to obtain only this limited information. Instead, data collection should (as the quotation from Kenny that we cited earlier urges) make every effort to measure all the variables that on theoretical grounds might be possible common causes of the variables of interest.

Our largely negative evaluation of the CLPC technique, we note, is shared by Cook and Campbell (1979). In a very reasoned appraisal of the technique, written by earlier proponents of the technique (1976), these authors state:

> We now see no advantage in reporting the results in the descriptive language of correlation coefficients if a plausible path model can be applied and the conclusions stated in path coefficient terms. [Cook and Campbell, 1979, p. 317].

We note that a plausible path model can always be constructed and the observed correlations interpreted in light of this model. It may not always be true that this model can be identified, but then this is an important fact to know in interpreting the correlations.

Appendix: CLPC as a Test for Spuriousness

As we stated in the text, it seems unlikely that CLPC will find wide application at the hands of those who grasp its extremely limited nature.

---

[7] There ain't no such thing as a free lunch. (See Heinlein, 1966.)

However, the analyst should be aware of the full range of conditions under which the test can prove useful. Some years ago, Kenny (1973) reported some results on the conditions under which differences of cross-lagged correlations vanish, but these results are unnecessarily restrictive. Here we report more general results concerning the validity of this test for spuriousness.

We omit proofs of our results in the interests of brevity, and because they are trivial to derive. Those interested in proofs should express the cross-lagged correlations in each model in terms of the structural parameters of the model, and then examine the consequences of imposing the stated equality constraints on the parameters.

We will be interested in exploring the ramifications of a variety of assumptions regarding the correlations of a set of n unmeasured Z_j with the observed variables X and Y. In the *stationarity* condition, we assume that $b_{X_t Z_{jt}}$ and $b_{Y_t Z_{jt}}$ do not depend on t; that is, they are constant over time. In the *quasi-stationarity* condition, we assume that these coefficients change proportionally over time; that is,

$$b_{X_1 Z_{j1}}/b_{Y_1 Z_{j1}} = b_{X_2 Z_{j2}}/b_{Y_2 Z_{j2}}.$$

In the *nonstationarity condition*, no constraints whatsoever are imposed on the regression coefficients linking X and Y with the Z_j. Next we vary the number of Z considered, treating explicitly the cases where $n = 1$ (only a single Z contributes) and $n > 1$. Third, we vary the constraint that X_t and Y_t are conditionally independent once the Z_{jt} are constrained. Finally, we consider the effect of measurement error on these results. Both the cases of constant reliability in measurement for X and Y, and the more general case of unconstrained reliability will be considered.

To simplify the notation, we define

$$b_{X_t Z_{jt}} = a_t,$$
$$b_{Y_t Z_{jt}} = b_t,$$
$$b_{Z_{j2} Z_{j1}} = c_j.$$

Prediction errors for X_t and Y_t will be denoted by e_{X_t} and e_{Y_t}. The correlations between these errors will be abbreviated as follows:

$$r_{e_{x_1} e_{y_1}} = e_1,$$
$$r_{e_{x_2} e_{y_2}} = e_2.$$

We will find it convenient to refer to those results that are obtained under a variety of conditions by their numbers in the following list:

1. $r_{X_1 Y_2} = r_{Y_1 X_2}$.
2. $r_{X_1 Y_2} r_{Y_1 X_2} = r_{X_1 X_2} r_{Y_1 Y_2}$.
3. No equality condition holds.
4. No individual parameters are identified.

I. Stationarity ($a_1 = a_2$, $b_1 = b_2$)
  A. Perfect Measurement
    1. If $n = 1$ and $e_t = 0$, then Results 1 and 2 hold. In addition,

$$c = r_{X_1Y_2}/r_{X_1Y_1} = r_{X_1Y_2}/r_{X_2Y_2} = r_{Y_1X_2}/r_{X_1Y_1} = r_{Y_1X_2}/r_{X_2Y_2},$$
$$a_t = r_{X_1X_2}/c,$$
$$b_t = r_{Y_1Y_2}/c.$$

    2. If $n > 1$ and $e_t = 0$, then Results 1, 2, and 4 hold.
    3. If $n = 1$ and $e_t \neq 0$, then Results 1, 2, and 4 hold.
    4. If $n > 1$ and $e_t = 0$, then Results 1, 2, and 4 hold.
  B. Stationary Measurement Error
    1. If $n = 1$ and $e_t = 0$, then Results 1 and 2 hold. In addition, $c =$ the four estimates given in I.A.1. No other individual parameters are identified.
    2. If $n > 1$ and $e_t = 0$, then Results 1, 2, and 4 hold.
    3. If $n = 1$ and $e_t \neq 0$, then Results 1, 2, and 4 hold.
    4. If $n > 1$ and $e_t \neq 0$, then Results 1, 2, and 4 hold.
  C. Unconstrained Measurement Error
    1. If $n = 1$ and $e_t = 0$, then Result 2 holds. In addition,

$$c = \{r_{X_1X_2}r_{Y_1Y_2}/r_{X_1Y_1}r_{X_2Y_2}\}^{1/2} = \{r_{X_1Y_2}r_{Y_1X_2}/r_{X_1Y_1}r_{X_2Y_2}\}^{1/2}.$$

No other individual parameters are identified.
    2. If $n > 1$ and $e_t = 0$, then Results 3 and 4 hold.
    3. If $n = 1$ and $e_t \neq 0$, then Results 3 and 4 hold.
    4. If $n > 1$ and $e_t \neq 0$, then Results 3 and 4 hold.
II. Quasi-Stationarity ($a_1/b_1 = a_2/b_2$)
  A. Perfect Measurement
    1. If $n = 1$ and $e_t = 0$, then Results 1 and 2 hold. In addition,

$$c = r_{X_1Y_2}/\{r_{X_1Y_1}r_{X_2Y_2}\}^{1/2} = r_{Y_1X_2}/[r_{X_1Y_1}r_{X_2Y_2}]^{1/2}$$
$$= \{r_{X_1X_2}r_{Y_1Y_2}\}^{1/2}/\{r_{X_1Y_1}r_{X_2Y_2}\}^{1/2},$$
$$a_1 = (r_{X_1Y_1}r_{X_1X_2}/r_{Y_1X_2})^{1/2}.$$
a_2, b_1, and b_2 can be solved by symmetry.
    2. If $n > 1$ and $e_t = 0$, then Results 1, 2, and 4 hold.
    3. If $n = 1$ and $e_t \neq 0$, then Results 1, 2, and 4 hold.
    4. If $n > 1$ and $e_t \neq 0$, then Results 1, 2, and 4 hold.
  B. Stationary Measurement Error
    1. If $n = 1$ and $e_t = 0$, then Results 1 and 2 hold. In addition, $c =$ the three estimates in II.A.I. No other individual parameters are identified.
    2. If $n > 1$ and $e_t = 0$, then Results 1, 2, and 4 hold.
    3. If $n = 1$ and $e_t \neq 0$, then Results 1, 2, and 4 hold.
    4. If $n > 1$ and $e_t \neq 0$, then Results 1, 2, and 4 hold.

C. Unconstrained Measurement Error
  1. If $n = 1$ and $e_t = 0$, then Result 2 holds. In addition,

$$c = (r_{X_1Y_2}r_{Y_1X_2}/r_{X_1Y_1}r_{X_2Y_2})^{1/2} = (r_{X_1X_2}r_{Y_1Y_2}/r_{X_1Y_1}r_{X_2Y_2})^{1/2}.$$

  No other individual parameters are identified.
  2. If $n > 1$ and $e_t = 0$, then Results 3 and 4 hold.
  3. If $n = 1$ and $e_t \neq 0$, then Results 3 and 4 hold.
  4. If $n > 1$ and $e_t \neq 0$, then Results 3 and 4 hold.

III. Nonstationarity
  A. Perfect Measurement
    1. If $n = 1$ and $e_t = 0$, then Results 1 and 2 hold. In addition,

$$c = (r_{X_1X_2}r_{Y_1Y_2}/r_{X_1Y_1}r_{X_2Y_2})^{1/2} = (r_{X_1Y_2}r_{Y_1X_2}/r_{X_1Y_1}r_{X_2Y_2})^{1/2},$$
$$a_1 = (r_{X_1Y_1}r_{X_1X_2}/r_{Y_1X_2})^{1/2}.$$

    a_2, b_1, and b_2 can be solved by symmetry.
    2. If $n > 1$ and $e_t = 0$, where the influences of only one of the Z_j are nonstationary, then Results 2 and 4 hold.
    3. If $n > 1$ and $e_t = 0$, then Results 3 and 4 hold.
    4. If $n = 1$ and $e_t \neq 0$, then Results 2 and 4 hold.
    5. If $n > 1$ and $e_t \neq 0$, where the influences of only one of the Z_j are nonstationary, then Results 2 and 4 hold.
    6. If $n > 1$ and $e_t \neq 0$, then Results 3 and 4 hold.
  B. Stationary Measurement Error
    1. If $n = 1$ and $e_t = 0$, then Result 2 holds. In addition, $c =$ the two estimates in III.A.I. No other individual parameters are identified.
    2. If $n > 1$ and $e_t = 0$, where the influences of only one of the Z_j are nonstationary, then Results 2 and 4 hold.
    3. If $n > 1$ and $e_t = 0$, then Results 3 and 4 hold.
    4. If $n = 1$ and $e_t \neq 0$, then Results 2 and 4 hold.
    5. If $n > 1$ and $e_t \neq 0$, where the influences of only one of the Z_j are nonstationary, then Results 2 and 4 hold.
    6. If $n > 1$ and $e_t \neq 0$, then Results 3 and 4 hold.
  C. Unconstrained Measurement Error
    1. If $n = 1$ and $e_t = 0$, then Result 2 holds. In addition, $c =$ the two estimates in III.A.I. No other individual parameters are identified.
    2. If $n > 1$ and $e_t = 0$, where the influences of one or more of the Z_j are nonstationary, then Results 3 and 4 hold.
    3. If $n = 1$ and $e_t \neq 0$, then Results 2 and 4 hold.
    4. If $n > 1$ and $e_t \neq 0$, where the influences of one or more of the Z_j are nonstationary, then Results 3 and 4 hold.

6

Separating the Effects of
Position and Change

It is sometimes meaningful to consider a model that contains both the static scores X_1 and X_2 and the change score ΔX as predictors of some outcome measure. For instance, in the mental health literature it is agreed that people's social origins (X_1) contribute to adult mental health to the extent that they help instill values and behavior predispositions that have implications for coping with stressful experiences thoughout life. Social destinations (X_2) are thought to contribute to mental health by determining the stresses to which people are exposed as adults. Mobility experiences (ΔX) are seen as distressing because they break stable social ties and lead to normlessness and anxiety (Kessler and Cleary, 1980).

There is a difficulty in estimating models of this sort. Since the change score is a linear function of the two static scores, it is impossible to obtain separate estimates for these three effects as long as they operate linearly. This problem was noted briefly in Chapter 2. In the present chapter we discuss it in more detail and review several approaches to its resolution.

We can begin by studying the basic algebraic difficulty. Consider the following linear equation.

$$Y = a + b_1X_1 + b_2X_2 + b_3\Delta X. \tag{6.1}$$

If we reparametrize this model in terms of the static scores, we have

$$Y = a + (b_1 - b_3)X_1 + (b_2 + b_3)X_2 \tag{6.2a}$$

or

$$Y = a + b_1^*X_1 \qquad + b_2^*X_2, \tag{6.2b}$$

which has only two structural parameters in three unknowns. This means that we cannot solve individually for the three parameters in Eq. (6.1).

Mathematically, it is impossible to separate these effects without making some additional assumptions. This impossibility conforms to the intuitive impossibility of an effect changing by (say) six units when we know that the two static scores differ by five units. Change is simply not something independent of the initial and final scores.

This problem is completely general to linear models even when they are expressed in terms of dummy variables. To see this, consider an example taken from Halaby and Sobel (1979). Suppose X takes on only five values at times 1 and 2. If we denote time 1 values as $X_{i.}$ and time 2 values as $X_{.j}$, we can treat the time 1 and time 2 static scores as dummy variables to predict an outcome Y by writing

$$Y_2 = a + b_1 X_{2.} + b_2 X_{3.} + b_3 X_{4.} + b_4 X_{5.} + b_5 X_{.2} + b_6 X_{.3} \quad (6.3)$$
$$+ b_7 X_{.4} + b_8 X_{.5}.$$

We have omitted terms in $X_{1.}$ and $X_{.1}$ from the equation to resolve trivial indeterminacies in the two sets of dummies. If we treat X as an interval-level variable, linear change effects will take the form of dummy variables X_{j-i}, in which each dummy represents a change of $j - i$ units from time 1 to time 2. However, it can readily be seen that

$$X_{(j-i)} = X_{.2} + 2X_{.3} + 4X_{.5} - X_{2.} - 2X_{3.} - 3X_{4.} - 4X_{5.}. \quad (6.4)$$

This equation shows that the same problem we encountered in Eq. (6.2) arises where linear change is studied with dummy variables: change effects cannot be distinguished from the effects of time 1 and time 2 static scores.

Given this difficulty of separating linear status and change effects, models to consider the independent influences of status and change are seldom seen in the social sciences. Yet, one occasionally comes upon a panel analysis in which it seems that causal processes conforming to a linear status-change effect model are at work.

A good illustration of this is found in the work of Eaton (1978), who studied the longitudinal relationship between exposure to a series of stressful life events (E_t) and scores on a self-report index of psychological distress (D_t) in a community sample interviewed first in 1967 and then again in 1969. His model was

$$D_2 = a + .55 D_1 - .16 E_1 + .28 E_2, \quad (6.5)$$

where the coefficients presented are standardized. All of these coefficients are significant at $p = .05$. Notice that the coefficient for E_1 is negative, something one would not predict on the basis of theory. It is counterintuitive to find that undesirable life experiences in the past will, after a period of time, *improve* one's mental health. However, as Eaton noted, this negative sign is easily interpretable as a change-score effect. While it makes no sense to claim that time 1 traumatic experience will directly improve time 2 mental health, it is very possible that a *change* for the better in one's situation over the time interval 1–2 can lead to improved mental health. If this change effect is large relative to the static score effect of E_1, the sign of the time 1 life event parameter will be negative. (This can be seen by referring back to question (6.2) and noting that the static-score and change-score parameters enter with opposite signs into the estimation equation.)

Unfortunately, if we assume that both E_1 and E_2 influence D_2 directly as well, it is impossible to disentangle the change-score parameter from the static-score parameters. Only when we impose some restriction on the static-score coefficients can we identify the influence of change.

There are three basic ways to impose a constraint of this sort. First, it might be possible to argue on substantive grounds that one of the static-score effects will be negligible. If so, then the change-score parameter can be identified by fixing this static-score parameter to zero (or some other specified value). If this is not plausible, we might make some restrictive assumption about the relative magnitudes of the two static-score effects. And finally, if this is not possible, a more complete empirical specification of the model in terms of causal processes intervening between the change score and the outcome might allow the separate influences to be identified.

In the remainder of this chapter we consider some illustrations of these various approaches.

Fixing Values of Parameters

The most obvious way to impose restrictions on the status-change model is to fix the value of one of the static-score parameters. The assumption usually made is that one of the static-score parameters is zero. But it could equally well be assumed that the parameter is some other fixed value.

In many cases in the literature this type of restriction is imposed in a seemingly naive way, without realizing that an identification issue is being resolved. Kaplan (1975), for instance, studied the influence of changing feelings of self-derogation on subsequent initiation of deviant behaviors in a sample of school children. The hypothesis was that children who begin feeling bad about themselves will have a tendency to become involved in new behavior patterns—some of them deviant—that allow them to regain positive self-feelings. To evaluate this hypothesis Kaplan estimated a model in which a dummy variable to represent initiation of various deviant behaviors between times 2 and 3 of a panel was regressed on a change score of the difference in self-derogation measured at times 1 and 2. No main effects for the static self-derogation scores at either time 1 or time 2 were included as predictors.

A model of this sort is fine if the researcher is willing to assume that the static scores are causally unimportant. In Kaplan's analysis, though, no justification is given for omitting these static scores from consideration.

In the Eaton data a justification of this sort is possible. Retrospective studies of the onset of psychological disorder (Brown and Harris, 1978) show that life events influence psychopathology only over a time interval of about one year. Eaton's two waves of data were collected 2 years apart, though, which means that the direct effect of E_1 on D_2 can be considered negligible.

Working with the assumption that E_1 has no direct effect on D_2, we can estimate the model

$$D_2 = a + b_1D_1 + b_2E_2 + b_3\Delta E \tag{6.6}$$

$$= a + b_1D_1 + b_2^*E_1 + b_3^*E_2, \tag{6.7}$$

where

$$b_2 = (b_2^* + b_3^*) \quad \text{and} \quad b_3 = -b_2^*.$$

When we reparametrize the Eaton model in these terms, we obtain $b_2 = .12$ and $b_3 = .16$. Here, as we would expect, both terms are positive, thus confirming Eaton's reasoning.

Constraining Relative Magnitudes of Parameters

At times it will be impossible to fix the value of either static-score parameter. When this is so one can still separate the effects of status and change if the relative magnitudes of the two static-score parameters can be constrained.

Hope (1975) originally suggested this approach to the problem. In his work on intergenerational mobility effects, he argued that the origin (X_1) and destination (X_2) statuses of an individual can reasonably be considered together as an overall status dimension. When this is done, change can be considered a second dimension. This assumes that both X_1 and X_2 influence the outcome of interest, but that they do so with equal magnitude.

Table 6.1
Values of the Change-Score Parameter b_2 for Different Values of p in the Eaton Data[a]

p	0	.1	.2	.3	.4	.5	.6	.7	.8	.9	1.0
b_2	.160	.171	.180	.188	.194	.200	.205	.209	.213	.217	.220

[a] All estimates of b_2 are significant at the .05 level. Estimates were obtained by solving for $b_2 = (b_1^* - pb_2^*)/(-1 - p)$, where $b_1^* = -.16$ and $b_2^* = .28$ are taken from Eq. (6.5).

We find the assumption of equal magnitude effects to be somewhat implausible in the published applications of the approach.[1] Nonetheless, it is conceivable that a strong case for consistency can be made in some instances. When the equal magnitude assumption is made we have

$$Y = a + b_1(X_1 + X_2) + b_2\Delta X. \qquad (6.8)$$

This model, like that in Eq. (5.1), is a reparametrization of Eq. (6.2b), where

$$b_1 = (b_1^* + b_2^*)/2 \quad \text{and} \quad b_2 = (b_2^* - b_1^*)/2.$$

When the assumption is that X_1 has a direct effect p times as large as that of X_2, we have

$$Y = a + b_1(pX_1 + X_2) + b_2\Delta X. \qquad (6.9)$$

And, in terms of the Eq. (6.2b) parameters, we have

$$b_1 = (b_1^* + b_2^*)/(1 + p)$$

and

$$b_2 = (b_1^* - pb_2^*)/(-1 - p).$$

The Eaton data could be studied by making a consistency constraint of this sort. Using the parameter values in Eq. (6.5) and varying p between 0 and 1, the change score coefficient b_2 takes on the values shown in Table 6.1. We see here that the change-score coefficient is positive and significant for all values of p between 0 and 1. It ranges from a minimum of .16

[1] For instance, Zurcher and Wilson (1979) study inconsistency effects among Navy Reservists by assigning equal main effect weights to the Navy Reserve rank and the civilian occupation status of the reservists. This assumption about the relative causal influences of status is totally arbitrary. It is perfectly conceivable, for instance, that civilian occupation has a substantial main effect on satisfaction with the Reserve, while Reserve rank has no effect; or vice versa. In the absence of any argument to support the choice of equal a priori weighting (the authors themselves present none), it is difficult to assess the validity of this assumption.

when $p = 0$, to a maximum of .22 when $p = 1$. Since it is implausible to assume that p would be less than zero or greater than one, these values represent bounds on the probable value of the true change-score effect.

We do not make an argument that any particular value of p is definitely the correct one. Indeed, it will seldom be the case that prior knowledge is sufficient to allow a precise argument of this sort. However, as this example demonstrates, it is possible in some cases to bound the values the change score can have within a narrow range.

Elaborating the Causal Model

The need to constrain the static score parameters can be avoided, at times, if other variables are brought into consideration. There must be good theoretical reason for choosing to take this approach, though, for the parameter estimates are highly sensitive to the particular specification used.

One way to do this is shown in Figure 6.1. We posit the existence of two latent intervening variables, "status" and "change." In the special case where the slopes from status to X_1 and X_2 are fixed at 1.0 and the slopes from change to X_1 and X_2 are fixed at -1.0 and 1.0 respectively, and where the error terms of status and change are fixed at 0 and no other predictors of status and change exist, this reduces to the model in Eq. (6.8). The parameters connecting the X_t to the latent variables are just-identified. Although we have included two Y scores in the figure, this remains true no matter how many Y are considered.

If substantive considerations suggest the existence of some exogenous Z, which is related to X_1 and X_2 only as a status effect, not a change effect, it is possible to identify the slope from the latent status variable to one of

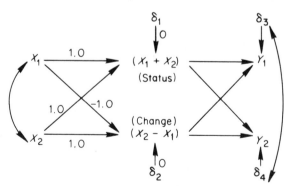

Figure 6.1. A single-indicator model of linear status-change effects.

the X_t. This model becomes progressively more overidentified as we include additional Z_k or Y_j in the model.

A second way to use additional variables to estimate the extent of consistency in the static score influences is shown in Figure 6.2. Here we have included observed indicators of the latent status construct. When empirical indicators of this sort can be included in one's model, it is possible either to conceptualize the latent status construct as an unmeasured factor or as a linear composite of the exogenous variables. In either case, as long as it is assumed that the latent change construct has no direct causal impact on the indicators, it is possible to identify the slope of status on X_2. As the number of indicators increases, this model becomes progressively more overidentified.

In general, models of this sort require some empirical determinant, effect, or indicator of one latent construct to be measured and assumed not to have a parallel relationship to the other construct. Whenever theory specifies a nonparallel relationship of this sort, it will be possible to separate the influences of status and change.

Nonlinear Status-Change Models

We mentioned previously that the embeddedness problem is a problem of *linear* models, since it is here that a function of a difference score will be identical to the difference of that function of the two static scores. When one is working with a nonlinear model this problem will not arise.

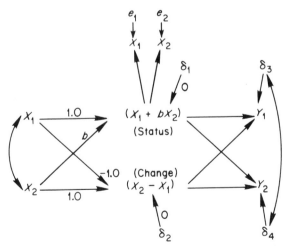

Figure 6.2. A multiple-indicator model of linear status-change effects. The parameter b represents the relative weight of X_2 as compared with X_1 in the definition of status.

To demonstrate this point we consider a model that is quadratic in the two static scores, X_1 and X_2, and also in an independent effect of the change score. That is,

$$Y = a + b_1X_1 + b_2X_1^2 + b_3X_2 + b_4X_2^2 + b_5\Delta X + b_6(\Delta X)^2. \quad (6.9)$$

Reexpressing this equation in terms of static scores, we have

$$\begin{aligned} Y = a &+ (b_1 - b_5)X_1 + (b_2 + b_6)X_1^2 + (b_3 + b_5)X_2 + (b_4 + b_6)X_2^2 \\ &- 2b_6X_1X_2 \end{aligned} \quad (6.10a)$$

or

$$Y = a + b_1^*X_1 + b_2^*X_1^2 + b_3^*X_2 + b_4^*X_2^2 + b_5^*X_1X_2 \quad (6.10b)$$

where

$$b_6 = (-^1/_2)b_5^*, \quad b_4 = (b_4^* - b_6) \quad \text{and} \quad b_2 = (b_2^* - b_6).$$

The linear terms in this model (b_1, b_3, b_5) are not individually identified. As we see in Eq. (6.10b), though, the quadratic terms (b_2, b_4, b_6) are identified.

In general, when a model containing both static- and change-score effects is estimated, the nonlinear components will be identified. The linear components will remain confounded. When a model contains only nonlinear terms, no confounding will occur.

In practice, the most common sorts of models that contain both linear and nonlinear components are multiplicative in the predictors, including the special case in which the multiplicative term X_iX_j is defined so that $i = j$, in which case the term is quadratic. Higher order multiplicative terms are also used in some cases (Taylor and Hornung, 1979). In models of this sort, the linear parameters have no substantive interpretations themselves, but only help locate the bend points of the nonlinear prediction surface in the hyperspace defined by the X_t and Y. When the scales of X_t are only unique up to a linear transform (that is, when we are working with scales which have no true zero point) these locations are arbitrary, and it is possible to transform the X_t in such a way that the parameters b_1^* and b_2^* in Eq. (6.10b) vanish. (This fact is demonstrated in the Appendix to this chapter.) Consequently, the indeterminacy that remains in models of this sort is of no real importance.

Nonlinear Consistency Models

Although few sociological examples of status-change models can be found in the literature, there is good reason to believe that some combination of assumptions about equality of static score effects and nonlinearity of change-score effects will at times be required to identify a plausible model. Imagine, for instance, that interest centers on the effects of media messages on behavior predispositions. Over the time span of a short panel it will be reasonable to assume that exposure to messages at times 1 and 2 are equally important influences on current predispositions. But to the extent that change in exposure has an effect, it will probably be the case that the extent of consistency of exposure, rather than the direction of change when it occurs, will be the important variable. Inconsistency is an unsigned measure of change, at its lowest when change is zero and higher when the change score is greater or less than zero. This poses certain technical problems of estimation.

If we actually believe that the inflection point of the change score occurs at the value 0, we can take the absolute value of the change score to obtain a nonlinear measure of change. But if we want to estimate an inflection point, or if we want to investigate the possibility that inconsistency effects are not symmetric around the inflection point, we must estimate a model that is nonlinear in the static scores. Models of this sort will not be developed here, but are discussed in Southwood (1978).

Appendix: The Transformation and Interpretation of Nonlinear Models

In models containing nonlinear terms in X_t and in which X_t has no true zero point, it is possible to transform X_t in such a way that the parameter associated with its linear component vanishes. This is an important result for the interpretation of nonlinear models and is derived in this appendix.

The method used here is an extension of the work of Stimson, Carmines, and Zeller (1978). Consider the following model:

$$Y = a + b_1X_1 + b_2X_2 + b_3X_1^2 + b_4X_2^2 + b_5(\Delta X)^2 + b_6X_1X_2. \quad (6.A1)$$

Notice that we have omitted a linear term for ΔX since it is redundant with the linear terms for the two static scores.

Let

$$X_1 = T_1 + d_1, \quad (6.A2a)$$

$$X_2 = T_2 + d_2 \quad (6.A2b)$$

define linear transformations of the observed X_t.

Then

$$Y = (a + b_1d_1 + b_2d_2 + b_3d_1^2 + b_4d_2^2 + b_6d_1d_2)$$
$$+ (b_1 + 2b_3d_1 + b_6d_2)T_1 + (b_2 + 2b_4d_2 + b_6d_1)T_2 \quad (6.A3)$$
$$+ b_3T_1^2 + b_4T_2^2 + b_5(\Delta X)^2 + b_6T_1T_2.$$

The parameters d_1 and d_2 are unidentified here. They can consequently be defined arbitrarily and still not influence our estimates of the remaining parameters. In particular, we can choose values for d_1 and d_2 in such a way that the sums of products operating on T_1 and T_2 in Eqs. (6.A2a)–(6.A2b) vanish. This can be done by solving simultaneously the equations

$$b_1 + 2b_3d_1 + b_6d_2 = 0, \qquad (6.A4a)$$

$$b_2 + b_6d_1 + 2b_4d_2 = 0 \qquad (6.A4b)$$

for d_1 and d_2. This gives

$$d_1 = (b_2b_6 - 2b_1b_4)/(4b_3b_4 - b_6^2), \qquad (6.A5a)$$

$$d_2 = (b_2b_6 - 2b_2b_3)/(4b_3b_4 - b_6^2). \qquad (6.A5b)$$

Consequently, using Eqs. (6.A5a) and (6.A5b) to transform X_1 and X_2 we can obtain

$$Y = (a^*) + b_3T_1^2 + b_4T_2^2 + b_5(\Delta X)^2 + b_6T_1T_2 \qquad (6.A6)$$

where a^* is obtained by inserting the values for d_1 and d_2 into the first element of Eq. (6.A1). In the transformed T_1, T_2 surface, a^* represents the value of Y at the origin—its minimum/maximum value.

Notice, though, that while this transformation removes the indeterminacy associated with the main effects of X_1 and X_2 in Eq. (6.A1), it is still important to study the higher order terms carefully for possible indeterminacy. In Eq. (6.A6), for example, an indeterminacy remains because only 3 of the 4 remaining parameters are independent. This is so because $(\Delta X)^2$ can be expressed in terms of the other variables in the model. That is,

$$(\Delta X)^2 = (X_2 - X_1)^2 = X_1^2 + X_2^2 - 2X_1X_2.$$

Some constraint must, therefore, be imposed on this solution if it is to be identified. This applies equally if we reparametrize in terms of X rather than T.

If we assume that the equation can be expressed in terms of polynomials of X_1 and X_2, without an interaction between these two variables (e.g., when b_6 in Eq. [6.A1] is zero) the model is identified, and the main effects of X_1 and X_2 can be transformed away. A similar solution exists for any equation that is linear in the regression parameters and multiplicative in X_t as long as indeterminacies among the polynomials and interaction terms are resolved. The value of a^* will have to be obtained anew in each different model, though, by working through a substitution and solution of simultaneous equations like that described in Eqs. (6.A1)–(6.A5b).

7

Serially Correlated Error

Up to this point the models we have considered share the basic OLS (ordinary least squares) assumptions about the prediction error e_i: that var(e_i) does not depend on i and that cov(e_i, e_j) = 0 when i is different from j. Here we are adopting the notation of Chapter 2 in defining e_i as the error term in the prediction equation for X_i (see, for example, Eq. [2.3]). The second assumption is particularly vulnerable in the case of panel analysis because the prediction error terms associated with successive measurements of a single variable will frequently be correlated.

This type of correlated error is a common phenomenon in the analysis of time series data and is known as "serial correlation." Standard approaches for detecting its presence and for correcting parameter estimates for the bias it can create have been developed for time series problems. In this chapter we discuss this type of error as it applies to panel analysis and review ways of correcting for it. As we show in the following, somewhat different considerations arise when the problem is approached from a panel rather than time series perspective.

Sources of Serial Correlation

In practice, serially correlated errors can arise because of misspecification of the functional form of the relationship between X_t and X_{t+1}, and because of an omitted exogenous variable. In this section we briefly review how these factors give rise to serial correlation.

It is easy to see why misspecifying the functional form of a prediction equation will lead to serial correlation. If a linear form is being used when a curvilinear relationship actually exists, patterns of systematic under- and over-prediction will arise along different parts of the range. Since values in this range will contain a certain degree of stability across time, the error terms will be serially correlated.

When omitted variables exist, serial correlation becomes one consequence. Assuming that the omitted variable or variables have some consistency through time, the prediction errors at successive times will be related through their common dependence on the omitted influences. To see this, suppose that X and Z both cause Y, but that Z has no effect on X. A three-wave model embodying these assumptions is shown in Figure 7.1, and the corresponding true structural equations for Y are

$$Y_2 = b_1 Y_1 + b_2 X_2 + b_3 Z_2 + u_1, \tag{7.1a}$$

$$Y_3 = b_1 Y_2 + b_2 X_3 + b_3 Z_3 + u_2. \tag{7.1b}$$

For simplicity we have assumed stationarity in the regression coefficients.

If Z_2 and Z_3 are omitted from the analysis, the effective error term that will be estimated in practice will be

$$e_1 = b_3 Z_2 + u_1 \quad \text{and} \quad e_2 = b_3 Z_3 + u_2.$$

In the typical case, Z will be somewhat stable over time, and this will lead to a positive correlation between the observed error terms even though u_1 and u_2 are uncorrelated. In addition, the correlation between Y and Z will cause Y_1 to be correlated with e_2 and e_3 even when it is not correlated with u_1. and u_2.. The correlations among these error terms can be modeled as in Figure 7.2.

The existence of these correlations among the error terms means that Y_2, for example, can no longer be used to identify Eq. (7.1b). Since Y_2 is proportional to e_1, and e_1 is correlated with e_2, Y_2 will be proportional to

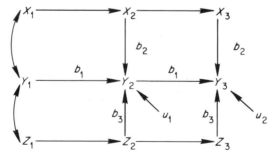

Figure 7.1. Omitted exogenous variable in a three-wave panel model.

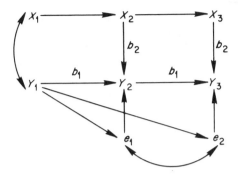

Figure 7.2. Serial correlation of errors in a panel model.

the error term for Y_3, contrary to the assumptions of OLS regression. A similar problem arises with regard to the use of Y_1 as an instrument to identify Y_2 or Y_3.

It is not difficult to show that failure to take this serial correlation into account will lead to biased estimates of b_1 and b_2. If Z is positively auto-correlated (as most exogenous variables are), a larger value for $cov(Y_t, Y_{t+1})$ will be found than in the absence of serial correlation. Examination of the normal equations derived from Eqs. (7.1a) and (7.1b) on the assumption that errors are uncorrelated (i.e., that there are no omitted exogenous variables) shows that this will inflate the estimate of the stability coefficient[1] b_1 and deflate the estimate for b_2, the effect of X on Y.

A third source of serial correlation among prediction errors is error in measurement. Whenever errors in the measurement of a given variable are related to measurements in the same variable at earlier or later times, serial correlation of errors will exist. As the treatment of measurement error is complex, we defer our discussion until Chapter 10, and assume here that all variables are measured without error.

Strategies for Dealing with Serially Correlated Errors

Since serial correlation is a common and serious problem in longitudinal studies it is natural to seek strategies for obtaining unbiased estimates in its presence. The strategy usually employed for doing this is gen-

---

[1] This entire discussion has been premised on the assumption that there is a lagged endogenous variable, and hence a nonvanishing stability coefficient in the true structural equation. For the reasons mentioned in Chapter 2, that will commonly be the case. Where it is not, parameter estimates will be unbiased even when serial correlation among errors is present, but OLS estimates will be reduced in precision.

eralized least squares (GLS), an extension of ordinary least squares in which the assumption that the variance–covariance matrix of prediction errors is diagonal, is replaced with the assumption that this matrix has a known, but not necessarily diagonal structure. When the assumed structure of the matrix is correct, and the prediction error matrix is fixed in the estimation procedure, GLS produces the best linear unbiased estimates of structural parameters.

The estimation of GLS equations is discussed in standard economic texts and will not be taken up here. Suffice it to say that estimation is not difficult once the researcher supplies fixed values for the off-diagonal elements in the prediction error matrix.

Coming up with estimates for the elements of the fixed matrix is a more difficult matter. Naively, one might think that one can estimate one's regression equations by neglecting correlations among the errors, and then computing the correlations among the errors obtained from the regressions. The trouble with this procedure is that if serial correlations are present but ignored in the regression, parameter estimates will be biased (as we just pointed out), and as a result, the estimates of error terms will also be biased. In addition, the correlations among these errors will be biased. Thus it is not a simple matter to generate an estimated fixed matrix empirically. And seldom will the researcher be able to derive specific nonzero values for the off-diagonal elements assumed to be serially correlated on the basis of theoretical considerations.

Several strategies have been devised in the time series literature to help the researcher come to some decision about the fixed matrix, some of which can be adapted for the analysis of panel data. But others cannot, thus making it difficult to adjust for serial correlation in panel models.

In time series applications the standard approaches for investigating serial correlation begin by studying the *form* of the error correlations. By "form" we mean the precise elements in the prediction error matrix that are serially correlated. This investigation is carried out by means of empirical correlogram analysis, a topic we take up in a different context in Chapter 8. Here, one estimates the residuals from an OLS analysis or an analysis making use of instrumental variables, finds the correlations among these residuals, and studies the pattern of correlated errors across a range of time intervals. Since theoretical analysis of the most plausible processes generating serial correlation shows that distinctive patterns in these correlations are produced when they are plotted against time, it is usually possible to infer the form of the serial correlation from an inspection of the plotted residual correlations.

This type of analysis makes use of the fact that time series data usually consist of at least several dozen observations for each variable. The same detailed analysis of time dependence is usually not possible when working with panel data because one typically has data for only a few waves. Since there is no other reliable method for estimating the form of serially correlated errors, it is consequently necessary to assume the form that these errors will take in panel analyses.

In the remainder of this chapter we review different assumptions one can make about the form of serial correlation and discuss how the magnitude of the serial correlation coefficients can be derived empirically once a particular form is assumed. As we stated above, once the form is fixed and the magnitudes of the nonzero elements in the prediction error matrix are derived, it is a simple matter to obtain a generalized least squares solution for the model.

Three different situations will be considered. In the first, we assume that the form of the serially correlated error is known even though we do not have any information about the underlying process responsible for the correlation. In the second, we assume that the time dependence of the omitted variables responsible for the serial correlation among prediction errors is known. In the third, we assume that no information is available about the form or the time dependence of the omitted variables.

Solutions Involving an Assumption about the Form

We saw earlier in this chapter that serial correlation of errors leads to a correlation between lagged endogenous variables and the prediction errors for the unlagged endogenous variable. When this occurs, none of the structural coefficients can be estimated without bias. However, if we are willing to make a specific assumption about the form of the serial correlation, it may be possible to transform the structural equation so that the correlation between the prediction error and the lagged endogenous variable disappears. Once this is done, OLS methods can be used to estimate the transformed equation without bias.

To illustrate the procedure, we consider the simplest possible case, one in which we have a lagged measure of Y and only one other predictor. That is,

$$Y_2 = a_1 Y_1 + a_2 X_2 + v_2, \tag{7.2}$$

where X_2 is an exogenous variable independent of v_2. We will assume that we know the form of the serial correlation to be "first-order autoregressive," which means that the prediction error depends on the error one time unit in the past. The "order" of the autoregressive term represents the number of lagged time units for which the errors are serially correlated.

The first-order autoregressive form can be written as follows:

$$v_t = \phi v_{t-1} + e_t. \tag{7.3}$$

Once this form is known it is possible to transform Eq. (7.2) so that the dependence between Y_1 and v_2 vanishes. This transformation makes use of a differencing procedure. First we lag Eq. (7.2) by one time unit to obtain

$$Y_1 = a_1 Y_0 + a_2 X_1 + v_1.$$

If this equation is multiplied by ϕ and the result subtracted from Eq. (7.2), we obtain

$$Y_2 - \phi Y_1 = a_1 Y_1 - a_1 \phi Y_0 + a_2 X_2 - a_2 \phi X_1 + v_2 - \phi v_1.$$

Bringing the term in ϕY_1 over to the right-hand side of the equation and making use of Eq. (7.3) to re-express the last two terms on the right, we have

$$Y_2 = (a_1 + \phi)Y_1 - a_1 \phi Y_0 + a_2 X_2 - a_2 \phi X_1 + e_2. \tag{7.4}$$

This procedure has removed the systematic component of the prediction error from Eq. (7.2) and has substituted in its place twice-lagged measures of Y and a single-lagged measure of X. This new equation can be estimated without bias using OLS methods.[2]

The value of the serial correlation coefficient ϕ can be estimated by obtaining the OLS coefficients and then deriving ϕ from them. For example, if we re-express Eq. (7.4) as

$$Y_2 = b_1 Y_1 + b_2 Y_0 + b_3 X_2 + b_4 X_1 + e_2,$$

we can solve for all the parameters by successive elimination, obtaining

$$a_2 = b_3, \quad a_1 = b_2 b_3 / b_4,$$

and two expressions for ϕ,

$$\phi = -b_4 / b_3$$

and

$$\phi = b_1 - b_2 b_3 / b_4.$$

---

[2] In this example, X_2 is exogenous to Y_2. But the method of dealing with serial correlation described for this example can also be employed when Eq. (7.2) is part of a simultaneous equation system that also includes an equation for the dependence of X_2 on Y_2 and X_1, with an error term μ_2 that is first-order serially correlated. After carrying out the subtraction procedure for both equations, estimation can proceed using simultaneous equation methods such as indirect least squares or two-stage least squares. Cross-lagged correlations of errors can also be handled in this manner.

The overidentifying condition on ϕ can be used as a consistency check for the model. Such overidentification is a general feature of regression equations that are obtained by a transformation that eliminates serial correlation.

Although the algebraic manipulations involved in checking for consistency are not difficult, there are times when one will want a single best estimate for ϕ and this the elimination procedure does not provide. There are nonlinear regression programs that use iterative procedures to find those parameters in Eq. (7.4) that will yield the smallest sum of squared prediction errors. The statistical properties of the parameters estimated in this way are not known and would probably be difficult to determine. Convergence of the iteration procedure is in general not guaranteed, and convergence to an incorrect estimate is possible, though these difficulties are less likely to occur if a plausible guess for the parameters is used to start the iteration (Hibbs, 1974). In any event, when a value of ϕ is obtained, either through algebraic manipulation or through nonlinear regression, the structural parameters can be efficiently estimated with GLS or by fixing the estimated value of ϕ in the prediction error covariance matrix of LISREL.

It is important to remember that this procedure for recovering the value of ϕ is entirely dependent on the assumptions embedded in the structural models. In the example we considered it was assumed that v_t was first-order autoregressive, that Y_0 had no direct effect on Y_2, and that the effect of X on Y was instantaneous, with no lagged component. If one or more of these assumptions is incorrect, the estimates of structural parameters will suffer from misspecification bias.

This situation cannot, in general, be remedied by trying out a variety of different specifications, since different assumptions about this type of model tend to lead to nothing more than different interpretations of long-lag parameter estimates. There are no purely empirical, totally atheoretical ways of distinguishing one type of serial correlation from other types in models containing only a few time points.

What is required, then, is thoughtful consideration of the reasons for expecting serial correlation to exist, and deductions from these considerations about the form these correlations are likely to take. If one assumes that serial correlation has its origins in the omission of exogenous variables, for instance, then the place to start is with the variable Z assumed to have been omitted. The simplest nontrivial assumption that can be made about Z is that it is first-order autoregressive:

$$Z_t = \lambda Z_{t-1} + \zeta.$$

If Z_t enters into Eq. (7.2) with a parameter a_3, then the nth-order serial correlations of the error terms of Y are $a_3^2 \lambda^n$. If one believes in the model, the correlations among the error terms can be used to identify the parameters a_3 and λ.

The special case where $\lambda = 1$ is worth noting. It corresponds to an omitted exogenous variable that remains constant with time. Subcultural influences that are expected to be extremely stable over the duration of a study might be an example. Here serial correlations of all orders are the same.

We see in these examples that the simplest and by no means empirically unrealistic assumptions about the time dependence of omitted exogenous variables lead to predictions that serial correlations will decline geometrically with lag, or that they will be constant, independent of lag. This is a very different assumption than the one usually made, namely that serial correlations are first-order autoregressive. In fact, Z would have to have a rather peculiar time dependence to produce error terms that have a non-vanishing first-order serial correlation, but vanishing serial correlations at all higher order lags.

Solutions Involving an Assumption about the Omitted Variable

It is clear from what we have just mentioned that while one can transform a structural equation to eliminate a postulated form of serial correlation, interpretation of the parameter estimates is uncertain if this postulated form is in doubt. It is clearly preferable to utilize information about the processes giving rise to serial correlation wherever possible. We outlined one method of proceeding where this type of information is available: to use the information about the omitted variables to determine the form of serial correlation expected, and then to transform the structural equation so as to eliminate the serial correlation.

If information about the time dependence of omitted exogenous variables is, in fact, available, it may at times be possible to eliminate the systematic component of the prediction error in another way. Consider the following model:

$$Y_2 = a_1 Y_1 + a_2 X_2 + a_3 Z_2 + v_2, \tag{7.5}$$

where the error term v_2 is serially uncorrelated. If the variable Z_2 is omitted from this equation, the new error term will absorb the Z_2 effect and become serially correlated.

Taking first differences gives us

$$Y_2 = (a_1 + 1)Y_1 - a_1 Y_0 + a_2(X_2 - X_1) + a_3(Z_2 - Z_1) + (v_2 - v_1). \tag{7.6}$$

The subtraction procedure is similar to what we did in the preceding section, but here we do not multiply the lagged equation by anything before we carry out the subtraction. Equation (7.6) contains a second-order autoregressive effect in Y even though none is present in Eq. (7.5), and it also contains terms in the lagged exogenous variables that were not present originally. In every instance, the coefficient of the lagged and instantaneous effect (or the first-lagged and second-lagged terms in the case of Y) are linearly related. In fact, the linear relationships are such that the exogenous variables contribute only in the difference.

If Eq. (7.6) is estimated on the assumption that Z_t does not contribute, then terms in $(Z_2 - Z_1)$ will contribute to the error term of the prediction equation. In fact, the error term will be

$$a_3(Z_1 - Z_0) + (v_2 - v_1).$$

This means that the covariance of consecutive errors (those separated by one time unit) will be

$$a_3^2 \text{cov}\{(Z_2 - Z_1), (Z_1 - Z_0)\} - \text{var}(v_1), \tag{7.6}$$

while for lags of k units, where k is greater than one, the corresponding covariance will be

$$a_3^2 \text{cov}\{(Z_{t+k} - Z_{t+k-1}), (Z_t - Z_{t-1})\}. \tag{7.7}$$

Inspection of Eqs. (7.6) and (7.7) shows that serial correlation will take certain known values when the time dependence of Z_t is given. In particular, when Z_t remains constant over time, the first term in Eq. (7.6) will vanish, and Eq. (7.7) will vanish entirely. Assuming that the v_t are homoscedastistic, the first-order serial correlation will be

$$-\sum_t v_t^2 / 2 \sum_t v_t^2 = -\tfrac{1}{2},$$

and serial correlations of all higher orders will vanish. Therefore, if we know that the omitted Z in our model remains constant over time, we can transform the original equation, which will have serial correlations of all orders, but of unknown magnitude, into a new equation by taking first differences. This new equation will have first-order serial correlation of $-\tfrac{1}{2}$, a known value, with all higher serial correlations of zero. The transformed equation can thus be estimated using GLS, assuming that we have first-order serial correlation with ϕ fixed at $-\tfrac{1}{2}$.

This procedure can be extended if Z is not constant, but changes in some known fashion. For example, if Z increases linearly with time, the systematic part of Z_t can be eliminated by taking differences twice instead of once. As long as the time dependence of Z can be expressed in a polynomial of finite order, it is possible to eliminate this systematic part of Z_t from the model with a finite number of subtractions.

Although our discussion of this transformation has implicitly assumed that X is exogenous to Y, the procedure is equally applicable when X and Y are jointly dependent endogenous variables (as in the models studied in Chapter 3) and Z affects both of them. The omission of Z in the estimation of a system of simultaneous equations will lead not only to serial correlation of errors in each of the variables separately, but also to correlations between the errors of the different variables, such as X and Y. Fortunately, whenever the equations for these variables are separately transformed to eliminate serial correlation or errors, the other correlations of errors will be eliminated as well.[3]

One caution is in order here; as much as in our previous discussion of methods for transforming serial correlations away, the procedure is very sensitive to misspecification. If an erroneous assumption is made about the temporal dependence of Z, serious errors in estimation may occur. For example, where Z does not remain constant as assumed, but changes in such a fashion that the difference $Z_2 - Z_1$ has the same sign as the dif-

---

[3] Suppose we have the following two simultaneous equations:

$$X_3 = a_1 X_2 + a_2 Y_3 + a_3 Z_3 + u_3,$$
$$Y_3 = b_1 Y_2 + b_2 X_3 + b_3 Z_3 + v_3,$$

and similar equations for X and Y at earlier times, with the same coefficients. Taking first differences and rearranging terms yields

$$X_3 = (1 + a_1)X_2 - a_1 X_1 + a_2(Y_3 - Y_2) + a_3(Z_3 - Z_2) + (u_3 - u_2),$$
$$Y_3 = (1 + b_1)Y_2 - b_1 Y_1 + b_2(X_3 - X_2) + b_3(Z_3 - Z_2) + (v_3 - v_2).$$

The presence of Z_3 in both untransformed equations leads to a covariance $a_3 b_3 Z_3^2$ between the errors of the equations for X_3 and Y_3 if the contribution of Z_3 is neglected in the estimation. For the differenced equation, the cross-sectional correlation of errors will be $a_3 b_3 (Z_3 - Z_2)^2$. Provided a_3 and b_3 are both different from zero, this will vanish only when $Z_2 = Z_3$, that is, when Z is constant over time. This is exactly the condition that will guarantee the vanishing of *serially* correlated error in both equations.

Nonvanishing covariances will also be found for the cross-lagged errors of X_t and Y_{t+1} in the untransformed equations, and all higher order lags as well. These covariances will be symmetric, in that the covariance for the errors of X_t and Y_{t+k} will be equal to that for the errors of Y_t and X_{t+k}. The transformation that eliminates the cross-sectional correlation of errors will also eliminate these cross-lagged correlations of errors.

When Z is not constant but increases linearly, an additional differencing will be needed to eliminate the cross-correlations of errors. But since the second differencing will also be needed to eliminate serial correlation of errors in this circumstance (see the text), no new complication is introduced. Any transformation that eliminates serial correlation entirely from one of the variables (or both) will also eliminate the cross-correlations of errors. It follows that the treatment of the text is just as applicable to reciprocal causal models as to unidirectional causal models.

ference $Z_1 - Z_0$, the first term of Eq. (7.6) will be positive and the correlation among the error terms will be less negative than $-\frac{1}{2}$. Its precise value will depend on the variances of the Z scores and the magnitude of the correlation between Z_t and Z_{t+1}. Furthermore, higher order serial correlations among the error terms will be present.

Solutions Involving Instrumental Variables

When no firm assumption about the form of the serially correlated error can be made, it is still possible to obtain consistent estimates of the parameters in panel models. This is achieved by using instrumental variables, which force the observed scores to be uncorrelated with the prediction error. The price paid for this consistency, though, is loss of precision for parameter estimates. In some cases this loss is substantial.

We start with a model of the following sort:

$$Y_2 = a_1 Y_1 + a_2 X_{21} + a_3 X_{22} + a_4 X_{23} + \cdots + a_n X_{2(n-1)} + v_2, \quad (7.8)$$

where X_{ij} denotes the value of X_j at time i, v_2 is uncorrelated with X_{ij} for all i and j, and serial correlation of unknown form is suspected among the errors.

Consistency can be achieved in a model of this sort by separating out from Y_1 the part known not to be correlated with v_t. This is easily done by regressing Y_1 on all the X_{2j} and all the X_{1j}. The result is a prediction equation consisting of $2(n - 1)$ exogenous predictor variables. Since v_2 is assumed to be uncorrelated with the X_j, the component of Y_1 systematically associated with (predicted by) these X_j will also be uncorrelated with v_2. Consequently, Eq. (7.8) can be estimated without bias by substituting the systematic component of Y_1 (the Y_1 predicted from the first-stage regression equation) for the observed Y_1. This estimation can be carried out with OLS. Since the procedure utilizes two regression analyses in sequence, the entire procedure is called "two-stage least squares." As the exogenous variables X_j serve as instruments for Y_1, the procedure is also an instrumental variables method.[4]

In time series analysis, this instrumental variable analysis is the first step in a procedure that goes on to generate empirical residuals from the

---

[4] It is necessary to include the X_{1j} as predictors of Y_1 in the first stage of the regression because if Y_1 were to be estimated from the X_{2j} alone, the predicted values of Y_1 would be a linear combination of the variables already included as predictors in Eq. (7.8). Under this circumstance, individual parameters cannot be identified. The inclusion of X_{1j} in the first stage of the regression overcomes this problem, but only because the X_{1j} have no direct effect on Y_2. Parameter estimates will be unbiased only if this assumption is correct. As we have had occasion to remark before, TANSTAAFL (see Chapter 5, footnote 7).

OLS equation, finds the correlations among these residuals across a range of time intervals, evaluates the correlations to deduce the form of the serial correlation, and then reestimates the equations with GLS. This last step regains some of the precision lost by the use of instrumental variables in the earlier part of the procedure (Hibbs, 1974). In panel analysis, it is generally not possible to deduce the form of the serial correlation from residual analysis. Therefore, the precision lost by using instrumental variables will not be regained.

Since this loss of precision can be large, instrumental variables should be used only as a last resort. The approach will be most successful when the set of X_j in the prediction equation is strongly related to Y_{t-1}. However, it is critical that these X_i should be uncorrelated with v_t. In multi-equation models, where Y and some of the X_j are reciprocally related and where the serial correlation is produced by an omitted common cause, the X_j will not be appropriate instruments.

8

Initial Model Specification

The preceding chapters have been concerned with the use of panel analysis to estimate *given* structural equation models. In this chapter we take up a problem that arises in the course of deciding the structural equations to estimate, that is, in initial model specification. In particular, we concentrate on methods for determining the length of the lagged relationship between two variables in a model.

The issues involved here are far from inconsequential. This is obviously true for deciding whether or not a causal relationship between X and Y should be estimated at all. But it is equally true for the issue of specifying the lag over which a presumed effect will be estimated. If one variable responds very rapidly to change in another variable, a model that assumes influences occur only after a long time may fail to detect them. On the other hand, if change occurs only very slowly, a model in which lagged effects are assumed to make their impact within a very short time span may fail to detect any significant change. As we show in the next chapter, even when change is detected, an erroneous assumption regarding the duration of the lag over which this change is felt may lead to parameter estimates that have the wrong sign and are badly biased in magnitude.

The decision to include the causal influence of one variable on another can usually be made on theoretical grounds; less often is it possible to decide on an appropriate causal lag on this basis, for almost never does social science theory make concrete predictions about the duration of causal effects. Ordinarily, the most we have are commonsense notions about the rough magnitude of time a given process is likely to take. We know, for example, that moods are susceptible to rapid swings, and consequently might look for the influence of another variable on mood with a time lag of hours or days. On the other hand, a lag this short would be quite implausible if we were studying the effect of unemployment on fertility. A minimum lag of 9 months would be required.

Given the importance of correct lag specification and the inability of theory to provide this specification, one might think it advisable to adopt an empiricist strategy by sampling observations over closely spaced time intervals and including *all possible* causal lags in the initial model. As long as this model is identified, the data will determine the appropriate causal lags.

The attraction of this strategy is that it requires no *a priori* information about the structure of the causal lags in the model outside of some general notions about the range of plausible time intervals over which a causal effect is likely to occur. Its drawback is that if multicollinearity among predictor variables is present (as it might well be for a relatively stable variable repeatedly measured at several closely spaced points in time) effects with different lags will be estimated with low precision when they are all included in the model. In fact, it can happen that an effect with a unique lag will appear to be spread out over a number of different lags, with none of the individual contributions achieving statistical significance (Gordon, 1968). This will complicate the process of respecifying an initially inadequate model and carrying out a new estimation of the respecified model to obtain an improved fit to the data.

To avoid these difficulties, it is advisable to include only a single lagged effect in one's initial model, but to use empirical information to determine the most appropriate length for this lag. After this initial model is estimated, it will still be possible to investigate different distributed lag specifications. But it is useful, as a first step, to begin with a single lag. We develop an approach for making such a determination below. As we show, it is a far from trivial matter to recover a true causal lag from the data. However, in panels collected over a moderate number of waves, restrictions can sometimes be placed on the possible length of the true causal lag.

The General Approach

We draw on two separate traditions of research for the approach taken in this chapter, one in applied time series analysis and the other in theoretical panel analysis. Both are based on the analysis of cross-correlograms, diagrams that graph the changing values of correlations between X and Y as time lags are systematically varied. These cross-correlograms are used as diagnostic devices to infer the temporal lag in a causal relationship between X and Y.

In the most simple formulation, we would expect that the series

$$r_{X_1Y_n}, r_{X_2Y_n}, \cdots r_{X_nY_n}, r_{X_{n+1}Y_n}, r_{X_{n+2}Y_n}, \cdots r_{X_{n+m}Y_n}$$

would have a single maximum value and that the lag corresponding to this value would identify the true causal lag for the dominant causal influence. It should be obvious to readers who have digested the material presented in Chapter 5 that this simple formulation bears much in common with the cross-lagged panel correlation (CLPC) technique. Among the points of resemblance is the fact that in both cases differential stability in the two variables can lead to erroneous conclusions. As we demonstrate below, the true causal lag does not always correspond to the maximum value in a cross-correlogram.

Efforts to overcome this difficulty in time series analysis have consisted of devising methods for removing the differential stabilities from the two series before estimating the cross-correlogram. These methods do not generalize readily to panel applications. And as a result, work on cross-correlograms has not progressed as far in the panel analysis literature as in time series applications. Also, the work that has been done to date has concentrated more on the analysis of theoretical correlograms than on empirical strategies for purging the data of interference from stability. This difference stems from the fact that panel data usually contain only a few degrees of freedom associated with time, and so cannot be used to model processes of residualizing, like those used in time series analysis, where observations are typically available for at least several dozen time periods. As a result, the analysis of panel correlograms has proceeded on the assumption that stabilities must be left in the series. This obviously poses interpretive difficulties.

The foundation for the theoretical correlogram approach in panel analysis was laid some years ago by Pelz and his collaborators in a series of investigations of the correlations generated by different underlying true models (Pelz, Magliveras, and Lew, 1968; Pelz and Faith, 1971; 1973; Pelz and Lew, 1970; Faith, 1973). These investigators pursued two lines of research. In the first they examined the patterns of correlation generated by known panel models. In the second, they tried to develop procedures for recovering underlying panel models from the correlations to which the models gave rise. This work showed that certain types of underlying causal models generate relatively distinctive patterns of correlations.

To date, these results have not proved useful in practice because the diagnostic methods thus far developed fail to distinguish among several models that may be plausible. Basing our work partly on the empirical approach taken in the time series literature, and to a larger extent on the work of Pelz and associates, we derive some new analytical results in this chapter that are more general than those obtained previously. These results permit inferences to be drawn about the structure of causal models from the observed patterns in cross-correlograms. We also show that there are limits to how far one can go with this approach.

Theoretical Correlograms

In theoretical correlogram analysis we begin with an assumed underlying causal model and generate analytical formulas for the expressions of the cross-correlations in terms of the structural parameters. These formulas, if not too complex, can be studied to tell us what empirical correlograms generated by underlying models of the form considered will look like. We begin by considering a cross-lagged panel model for two variables.

The path diagram for this model is shown in Figure 8.1; the corresponding structural equations are

$$X_t = aX_{t-1} + dY_{t-1} + \epsilon_t, \tag{8.1}$$

$$Y_t = cX_{t-1} + bY_{t-1} + \nu_t, \tag{8.2}$$

and all variables are standardized.

When we have encountered models like this in the past, we have taken the cross-sectional correlations among the variables in the first wave of observations as given, and did not subject them to causal analysis.

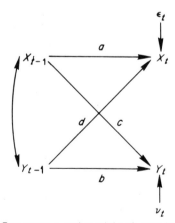

Figure 8.1. Two-wave panel model with cross-lagged effects.

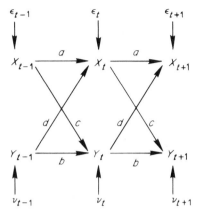

Figure 8.2. Stationary panel model with cross-lagged effects.

Although the same approach could be adopted here, it is much more convenient for our present purposes to assume that the structural equations that define our model have been in effect for a very long time with the *same* model parameters. In the vocabulary of Chapters 3 and 4, we regard this model as *stationary*. If we exclude from consideration systems in which the variables are exploding (getting larger and larger in magnitude over time, without approaching any asymptotic limit), the system will be arbitrarily close to equilibrium, and the correlation between any two variables will for all practical purposes depend only on the time lag between them, and not on time itself. This means that we can assume equalities like

$$r_{X_1 Y_1} = r_{X_2 Y_2}, \; r_{Y_1 Y_3} = r_{Y_2 Y_4},$$

and so forth. As we noted in Chapter 5, the assumption that systems are in equilibrium is not always valid, but our analysis will be limited to those cases where it is.[1]

Assuming stationarity for parameters, and that our system is in equilibrium, we use Eqs. (8.1) and (8.2) to derive expressions for correlations involving various lags. For the present, we assume that error terms are serially and cross-sectionally uncorrelated; later we investigate the consequences of abandoning these assumptions. The path diagram that embodies the assumptions of our model is displayed in Figure 8.2.

---

[1] A system may not be in equilibrium because its structural parameters are not stationary, or, when its parameters are stationary, because insufficient time has elapsed for transient effects to die out. One can check on whether equilibrium has been reached by testing for the statistical significance of the difference between adjacent correlations.

To avoid overly cumbersome notation, we define

$$s_n = r_{X_t X_{t-n}}, \tag{8.3a}$$

$$t_n = r_{Y_t Y_{t-n}}, \tag{8.3b}$$

$$r_n = r_{Y_t X_{t-n}}, \tag{8.3c}$$

$$r_{-n}^* = r_{X_t Y_{t-n}}. \tag{8.3d}$$

These equations are defined for all integral values of n, but $s_n = s_{-n}$ and $t_n = t_{-n}$, so we can confine our attention to nonnegative values of n for these quantities. It is not necessarily the case that $r_n = r_{-n}$, but since $r_n = r_{-n}^*$, we can restrict our attention to nonnegative values of n in computing these quantities as well.

In the abbreviated notation of Eqs. (8.3a)–(8.3d), the normal equations corresponding to the structural Eqs. (8.1) and (8.2) become

$$s_n = as_{n-1} + dr_{n-1} \tag{8.4a}$$

$$r_n = br_{n-1} + cs_{n-1} \tag{8.4b}$$

$$\left. \begin{array}{c} \end{array} \right\} \; n \geq 1.$$

$$r_n^* = ar_{n-1}^* + dt_{n-1} \tag{8.4c}$$

$$t_n = bt_{n-1} + cr_{n-1}^* \tag{8.4d}$$

These recurrence relations express the autocorrelations (s_n and t_n) and cross-correlations (r_n and r_n^*) lagged by n time units in terms of those lagged by $n - 1$ time units. By using these equations, one can generate all the autocorrelations and cross-correlations sequentially, beginning with the unlagged correlations ($n = 0$). Our two-equation model generates four correlograms, one for each of the sequences s_n, r_n, r_n^*, and t_n. In our graphical presentation, though, we will collapse r_n and r_n^* into a single sequence with negative and positive values of n by making use of the equality of r_{-n} and r_n^*.

Analytical work on the properties of these sequences has not been pursued very far because the expressions generated by this procedure are complex and their derivation is tedious. The correlations must be computed one at a time, beginning with expressions[2] for s_0, t_0, r_0, and r_0^*, and

---

[2] To generate correlations from the recurrence relations we need the correlations with zero lag. The zero-lagged autocorrelations s_0 and t_0 are, of course, equal to one. To obtain $r_0 = r_0^*$, we multiply Eq. (8.1) by Eq. (8.2), sum over cases, and make use of the stationarity condition to obtain

$$r_0 = abr_0 + cdr_0 + ac + bd.$$

Solving for r_0, we have

then using the recurrence relations of Eqs. (8.4a)–(8.4d) to derive s_1, t_1, r_1, and r_1^*. From these, s_2, t_2, r_2, and r_2^* can be computed in turn. Continuing in this fashion, expressions can be derived for all the correlations in each correlogram. It is evident that this is a laborious procedure.

It is at this step that our work takes a different approach than that adopted by the Pelz group. Rather than work with these exceedingly cumbersome expressions, we have derived summary expressions for the entire sequence of correlations all at once. This derivation is difficult, but results in a tractable expression. The derivation, and a detailed discussion of the results obtained by analyzing the expressions it yields, are presented in a technical appendix to this chapter. Here we merely state the most general expression without proof. We find that r_n takes the following form:

$$ r_n = C_1 m_1^n + C_2 m_2^n, $$

where C_1 and C_2 are arbitrary constants and m_1 and m_2 are the two solutions to the quadratic equation

$$ m^2 - (a + b)m + (ab - cd) = 0. $$

It can be shown that r_n^*, s_n, and t_n all obey this same equation with the same coefficients.

The behavior of the correlograms and cross-correlograms as n varies depends on the character of the roots m_1 and m_2. Depending on the values, the cross-correlograms can oscillate with or without damping, decline monotonically to a given asymptote, or rise to a peak and then decline. Although there is some ambiguity as to the sorts of causal processes

$$ r_0 = \frac{ac + bd}{1 - ab - cd}. $$

If we make use of the series expansion (valid for $|x| < 1$)

$$ 1/(1 - x) = 1 + x + x^2 + \ldots, $$

we can expand this expression for r_0 as follows:

$$ r_0 = (ac + bd)\{1 + (ab + cd) + (ab + cd)^2 + \ldots\}. $$

Each term in this expansion represents the contribution of a given path in the path diagram. For example, the term ac represents the contribution of the path in Figure 7.2 that goes backward from X_{t+1} to X_t and then forward to Y_{t+1}; the term bd represents the contribution of the path that goes backward from X_{t+1} to Y_t and then forward to Y_{t+1}. Terms involving higher powers of the parameters involve longer lags. For example, the term ac^2d arises from the path that goes backward from X_{t+1} to Y_t, continues backward to X_{t-1}, then goes forward from X_{t-1} to X_t and then to Y_{t+1}.

that might be responsible for each of these patterns, a good deal can be said about the presence or absence of the cross-coefficients c and d in the underlying causal models, and about their signs, from an inspection of these patterns.

We turn now to a more detailed consideration of these patterns, and the inferences that can be drawn from them.

Basic Patterns of Correlograms

We consider four possibilities: c and d are both zero; either c or d is zero, but not both; c and d are both nonzero and of equal sign; c and d are both nonzero and of opposite sign. After deducing the patterns of cross-correlation generated by each of these different possibilities we draw some diagnostic inferences.

(a) Both Cross-Effects Are Zero. As long as the errors of X and Y are uncorrelated, as we have assumed them to be, all cross-correlations in the sequences r_n and r_n^* will be zero (this will be true even if the system is not in equilibrium). Unfortunately, the diagnostic implications of this fact are not large, since it will often happen that correlations in the two sequences will be nonzero because of omitted common causes of X and Y. The time dependence of the two sequences are then indeterminate without some information about the omitted common causes.

(b) One Cross-Effect Is Zero. For definiteness we assume that $d = 0$. This corresponds to the assumption that Y has no influence on X, but X exerts an influence on Y of unknown magnitude c. The expressions for the correlograms simplify in this case, taking the following form:[3]

$$s_n = a^n \tag{8.5a}$$

$$t_n = b^n + \frac{ac^2}{1 - ab} \frac{(b^n - a^n)}{b - a} \tag{8.5b}$$

$$r_n = \frac{c}{(1 - ab)(b - a)} [b^n(1 - a^2) - a^n(1 - ab)] \tag{8.5c}$$

$$r_n^* = a^{n+1}c/(1 - ab) \tag{8.5d}$$

$n \geq 0.$

---

[3] Pelz and his collaborators obtained these results by applying the second of the two methods described in the Appendix to the case where $d = 0$. They did not, however, develop analytical expressions for the more general case.

Here the sequence r_n^* declines geometrically[4] from its peak at $n = 0$, the height of the cross-correlogram increasing with c and with the stability coefficients a and b. The behavior of the sequence r_n is more complicated. Intuitively, we might expect that the magnitude of r_n would be largest at $n = 1$, the lag we have assumed to be the causal lag in our model. However, as Pelz and Lew (1970) have noted, and as can be verified readily from the expression for r_n in Eq. (8.5c), this is true only when a specific condition on the autoregressive parameters a and b holds, namely, that $(1 + a)b < 1$ (or, equivalently, that $(1 - b)/b > a$). When this inequality holds, and the parameter c is positive, r_n rises from $n = 0$, peaks at $n = 1$, and then declines to a limit of zero at large n. When c is negative a mirror pattern is found: r_n falls from its value at $n = 0$, reaches a minimum at $n = 1$, and then rises to a limit of zero. In either case, all the r_n are of the same sign.[5]

When this inequality among the stability coefficients is violated, the peak of r_n is displaced from $n = 1$ to *higher* lags. If follows that the causal lag cannot always be inferred correctly from the location of the peak, because the true lag may not be as high as the lag for which the cross-correlation has the largest magnitude. However, the true lag *will never be higher* than the peak lag, since the displacement is always in the direction of higher lags. The location of the correlogram peak in this case places an upper bound on the true lag.

Figure 8.3 shows the region in the $(a-b)$ plane where peak displacement is expected. For a given value of a, displacement will occur whenever b is

---

[4] A departure from geometricity may suggest that d differs from zero. But caution is in order here; for a departure from geometricity can also originate in serial correlations among the error terms of the X_t, or in omitted variables. However, an omitted Z will contribute to a cross-correlogram only if Z influences both X and Y. This is a somewhat less likely eventuality than Z influencing X *or* Y, but not both. Thus a geometric pattern of decline will tell us that d is zero, but a departure from this pattern will not tell us for certain that d is different from zero, unless other information is available to exclude alternative explanations for the departure.

[5] Setting $c = 0$ in Eq. (8.5b), we also see that both s_n and t_n decline geometrically from their peak values at $n = 0$. However, if the autoregressive influences among the X_t and/or Y_t are not simplex (that is, if higher order autoregressive effects are present), the decline will not be strictly geometric. When c differs from zero, s_n, the autocorrelations of X_t, continue to decline geometrically from $s_0 = 1$. Thus any departure from geometric decline among the s_n indicates that $d \neq 0$. Similarly, when $c \neq 0$, the t_n decline from $n = 0$, but not necessarily geometrically. Thus one can test for the presence of cross-effects by examining the auto-correlograms. This, however, is not a very stringent test for the presence of cross-effects, since a departure from a pattern of geometric decline in one of the autocorrelograms can have other causes, including higher order autoregressive effects and omitted variables that influence one of the included variables. We have specified the model so as to exclude these possibilities, but in practice it is not always possible to rule them out.

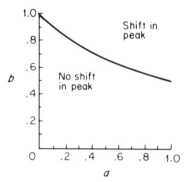

Figure 8.3. Regions where shift in cross-correlogram peak is, and is not expected.

large enough—and the larger b, the larger the number of lags by which the peak will be displaced. Although the value of b is not known at the preliminary stage of data analysis, high autocorrelations among the Y_t for those lags that are comparable to the time scale of the causal processes at work are a good indication that b is large, and that displacement is thus a real possibility.

Fortunately, it is precisely when the autocorrelations of Y_t are extremely high that misspecification of the causal lag should do the least harm, for if Y_t and Y_{t-n} are highly correlated, then it will make little difference which is used in estimating the model (this assertion is documented in Chapter 9). The greatest danger of temporal misspecification will be present when b is high enough to displace the correlogram peak, but not so high as to eliminate specification bias.

(c) *Cross-Effects with Different Signs.* When both c and d exist, but have opposite signs, either of two patterns may appear in the cross-correlogram. When the quantity $(a - b)^2 + 4cd$ is negative, oscillatory behavior will appear in the cross-correlograms. Under the defining conditions of the model we are considering (cross-lagged, first-order autoregressive, serially uncorrelated errors) this is the only case in which oscillatory behavior appears in a cross-correlogram.

A fuller understanding of the oscillatory behavior can be gained by reading the technical Appendix to this chapter. But the implications are easily grasped. Since the quantity $(a - b)^2$ is nonnegative, the entire quantity can be negative only if the second term, $4cd$, is negative and large enough in magnitude to outweigh the effect of the first term. This can happen only if both c and d are different from zero and are of opposite sign. A typical pattern of the oscillation that occurs in a cross-correlogram generated by such a pattern is shown in Figure 8.4.

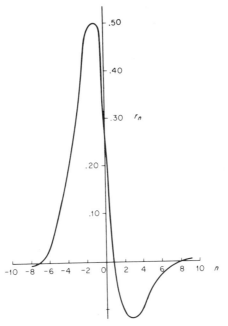

Figure 8.4. Oscillatory correlogram. In the figure, r_{-n} is defined to be equal to r_n^*. $X_t = .4X_{t-1} + .4Y_{t-1} + \epsilon_t. \ Y_t = .7Y_{t-1} - .2Y_{t-1} + \nu_t.$

If c and d are both different from zero and of opposite sign, but the quantity $4cd$ is smaller in magnitude than $(a - b)^2$, the correlograms will not oscillate. As long as the stability coefficients are at least moderately close to one another, only very small cross-effects will be missed by their failure to produce oscillation in the correlogram, but if a and b are far apart, even a cross-effect of moderate size can be missed by this criterion. However, as we will show below, another detection procedure can be used to identify a situation of this sort. Before this procedure can be described, though, we need to discuss the case in which both c and d have the same sign.

(d) Cross-Effects Have the Same Sign. In the case where c and d both differ from zero and are of the same sign, we may find any one of three possibilities: r_n and r_n^* both peak at lags that differ from zero; one has a peak while the other does not; neither has a peak at a lag other than zero, meaning that each peaks at $n = 0$ and declines monotonically with increasing n. To make matters even more complex, these same three possibilities exist for the case where c and d are of opposite sign, but $4cd$ is

smaller in magnitude than $(a - b)^2$. The first pattern, where r_n and r_n^* have distinct peaks at nonzero lags, is illustrated in Figure 8.5.

Here we see two major difficulties in drawing inferences about the structure of underlying models from correlograms alone. First, these patterns show that when both cross-effects are present peak displacement may occur upward or downward. The location of the peak in either cross-correlogram is, therefore, no longer an upper bound to the true lag. Second, the absence of a peak at nonzero lags is consistent with the possibility that the lagged variable in the cross-correlogram has a causal influence on the nonlagged variable. And, although the sign of the cross-correlogram (positive or negative) will usually correspond to the sign of the corresponding cross-effect (c or d), this will not always be the case. For particular sets of parameter values, the mutual influences of X and Y on one another can create lagged correlations of opposite signs to the true signs of the causal influence. The most that can be said from the correlograms alone is that the presence of peaks in both cross-correlograms shows both causal effects to be present.

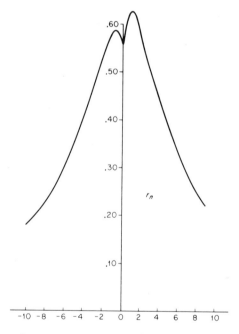

Figure 8.5. Nonoscillatory correlogram. In the figure, r_{-n} is defined to be equal to r_n^*. $X_t = .4X_{t-1} + .4Y_{t-1} + \epsilon_t$, $Y_t = .7Y_{t-1} + .2X_{t-1} + \nu_t$.

Diagnostic Implications

Although the preceding investigation showed that there is ambiguity in the interpretation of cross-correlograms, certain diagnostic guidelines can be drawn. These guidelines are based on the assumption that complications like omitted third variables, serially correlated error, and distributed lags (all complications not considered in Eqs. [8.1] and [8.2]) are absent, and that the process is stationary and in equilibrium.

Inferring the Presence of an Effect

If all the preceding assumptions are made, several rules can be stated for inferring the presence of cross-effects from cross-correlograms:

1. The presence of oscillatory behavior in the cross-correlograms shows that both cross-coefficients are present and of opposite sign.
2. The presence of nonzero peaks in both cross-correlograms (r_n and r_n^*) shows that both cross-effects are present, since we know that two peaks of this sort cannot be generated when $c = d = 0$; or when c or d is zero, but not both. No statement about the signs of c and d can be made in this case.
3. The presence of a nonzero peak in only one cross-correlogram in a pair can be generated by a variety of different underlying structures. In all of these, though, *at least* one of the coefficients c and d must be nonzero.
4. The absense of nonzero peaks in either cross-correlogram provides no diagnostic information. This case is consistent with the possibility that both cross-effects are zero, that one is zero while the other has an effect with a short temporal lag, or that both effects are nonzero.

Inferring the Temporal Lag of an Effect

When the analyst is confident that one of the cross-effects is zero, it is possible to place bounds on the true causal lag of the other cross-effect. The rule here is that the peak in the lagged variable's cross-correlogram is an *upper bound* on the true causal lag. The lower bound, of course, is zero.

The lag displacement is much more sensitive to the stability of the endogenous variable than the exogenous variable. When the former is no larger than .5 it is safe to assume that lag displacement has not occurred.

(Sensitivity analysis for the bias introduced by lag displacement is discussed in Chapter 8.)

If it cannot be assumed that one of the cross-effects is zero, no inference can be made about the true causal lag from cross-correlogram analysis. The peak can no longer be interpreted as an upper bound and the absence of a nonzero peak cannot be interpreted as evidence that the lagged effect is zero.

It is conceivable that, in some cases, a procedure similar to prewhitening in time series correlogram analysis will prove useful in these more complex cases. To infer correctly the upper bound on the true causal lag we do not need to destroy the stability in the two variables; we only need to know that causal influence is operating in only one direction. Therefore, *partial* cross-correlograms can be analyzed. To specify the upper bound of c (the effect of X on Y), for instance, we can examine the cross-correlogram r_n ($r_{Y_t X_{t-n}}$) controlling statistically for all earlier values of Y. This effectively purges the series of causal influences of Y on X.

There are two difficulties with this procedure. First, the partialling operation itself presents some problems. In order to control on *earlier* values of Y, it is necessary to consider at least one less value in the cross-correlogram than there are waves in the panel. If we are truly unsure about the duration of the lagged effect of Y, we would want to control even earlier values of Y. This means that the correlogram would be even more truncated.

This is not a major problem in time series analysis, where there are a good many time points. But it is a serious problem in panel analysis, where there are only a few degrees of freedom associated with time. In a six-wave panel, for instance, it is possible to estimate the cross-correlogram series r_5, r_4, r_3, r_2, r_1, r_0. This might well be long enough to give the analyst firm information about the upper bound lag of c, assuming that d is zero. However, if the magnitude of d is unknown and we are confident only that its lag is between 0 and 4, then it is only possible to estimate the *partial* cross-correlogram series r_1, r_0. This pair of correlations is totally inadequate for any diagnostic purposes.

Even when a long series exists, or when we are confident that the true lag of d is short (which means that we only have to partial on one or two earlier lags of Y in the partial cross-correlogram), a second difficulty arises when using a partial correlogram approach: whether or not to control Y_{t-n} when estimating the correlation $r_{Y_t X_{t-n}}$. If a *lower* bound on d of $n \geq 1$ cannot be assumed, then it is necessary to control Y_{t-n} to ensure that all influences of d have been controlled out of the series r_n. However, this means that r_0 no longer can be estimated as part of that series. Given the central role played by the cross-correlation at $n = 0$, this is a critical problem. Indeed, when a priori considerations cannot be used to justify omitting Y_{t-n} as a control variable, the partial cross-correlogram approach cannot be used successfully.

Extension to More Complex Models

So far our discussion has been predicated on the assumption that the lag for the $X \rightarrow Y$ equation is the same as the lag for the $Y \rightarrow X$ equation, that influences occur with only a single rather than a distributed lag, and that correlations among error terms are absent. In practice, one or more of these assumptions can be violated. We turn next to a discussion of these effects.

Unequal Lags

The great majority of panel studies implicitly assume that causal influences among variables all occur with the same lag. We know of no substantive reason to think that this will always be the case.

To investigate models with unequal lags, we may consider a two-variable model in which all autoregressive effects are first order, with lag one, but in which the effect of X on Y is lagged by k units, while the effect of Y on X is lagged by g units, with $k \geq 1$ and $g \geq 1$.

The same methods used in the appendix to derive the correlograms can be used in this case as well, but an analytical investigation of the solutions is more difficult because one must deal with the roots of higher order polynomial equations rather than quadratic equations. We have been able to verify for the case $k = 2$, $g = 1$ that oscillatory solutions are obtained when c and d are of the same or opposite sign, but have not analyzed other cases.

In a simulation study involving one cross-effect lagged at three time units and the other lagged at five units, Pelz, Magliveras, and Lew (1968) verified this behavior: damped oscillations were observed in both autocorrelograms and cross-correlograms whether c and d were both positive or of opposite sign. In the former case, all the correlations in a given correlogram were of the same sign; in the latter case, the correlograms changed sign as they oscillated. Although the first (and highest) peaks in both cross-correlograms occurred at the right lags, the stability coefficients in the model studied were only moderately large (a situation unlikely to lead to displacement), and no investigation of other cases was reported.

Further investigation is clearly needed to determine how generally the properties found in these special cases hold. Our finding that the comparable properties did not necessarily hold in the equal lag case should make us suspicious that they may not always hold in the unequal lag case either. However, where oscillatory correlograms are found with first peaks at unequal lags, a model involving reciprocal causation with unequal lags should be considered, and the signs of the correlations in the correlograms may be taken as a clue regarding the signs of the cross-effects.

Distributed Lags

In all the models we have considered, the effect of X on later values of Y occurs with only a single lag. This assumes that X has little effect on Y until a time close to the lag length is reached, and stops having an effect shortly after that time. There are occasions when such an assumption may be valid. But if there are some variables that, like old soldiers, never die but just fade away, it may be more realistic to assume that the effect of X on Y is distributed over a number of lags. Models that incorporate such effects have been used widely in econometrics.

Two questions can be asked about such models: (*a*) can we distinguish distributed lag models from single-lag models through our preliminary diagnostic procedures, and (*b*) does it make much difference to our substantive findings if a true distributed lag model is estimated on the assumption that we have only a single lag?

Pelz, Magliveras, and Lew (1968) addressed the first of these questions empirically by computing correlograms for a two-variable panel model with unidirectional causation and varying assumptions about the distribution of lagged effects. Specifically, they found that correlograms were very similar for models with the same (moderately high) stability coefficients, whether it was assumed that only a single lagged effect was present, or whether lags followed a rectangular, quasi-normal, or geometric distribution, when the sum of lagged coefficients was held constant. We would expect the correlograms to coincide exactly when the sum of lagged coefficients remains the same as long as predictor variables are perfectly correlated; this empirical finding suggests that the correlograms will not be easily distinguished even when the stability coefficients are only moderately high. This dashes hopes of identifying distributed lag models from single models from the general shape of the correlogram.

An examination of the *partial* correlograms should prove more fruitful in identifying distributed lags; however, the simulations of Pelz, Magliveras, and Lew (1968) suggest that if one is interested in the long-run effect of one variable on another, rather than the precise time dependence of influence, the distribution of lags can be ignored.

The inclusion of lagged endogenous variables in one's model also makes explicit consideration of distributed lag effects less critical. Let us assume that lagged effects are geometrically distributed. At times, such a distribution is expected on theoretical grounds. For example, macroeconomic theory predicts that the effect of investment on income will take the form of a geometrically distributed lag. Even where theory makes no specific prediction, causal effects are often expected to die out gradually, and for these problems, the geometric distribution should be a reasonably good approximation. After exploring the consequences of assuming a geometric distribution, we will examine the effect of deviations from this distribution.

For the time being, then, we write

$$y_t = b(x_t + qx_{t-1} + q^2 x_{t-2} + \ldots) + u_t, \qquad (8.6)$$

where $0 < q < 1$, and the disturbance terms u_t are assumed to be serially uncorrelated. Lagging this equation by one time unit and multiplying the result by q, we have

$$qy_{t-1} = bq(x_{t-1} + qx_{t-2} + \ldots) + qu_{t-1}. \qquad (8.7)$$

Subtracting Eq. (8.7) from Eq. (8.6) and rearranging terms, we have

$$y_t = qy_{t-1} + bx_t + (u_t - qu_{t-1}). \qquad (8.8)$$

In place of an infinite number of cross-effects, we have an equation with only a single, instantaneous cross-effect, a lagged endogenous variable, and first-order serial correlation[6] among the error terms of $-q/(1 + q^2)$. We can estimate q and b from Eq. (8.8), and thus recover the original equation involving distributed lags, without ever having to estimate an equation containing distributed lags explicitly. The reader may wonder, though, how the researcher who estimates an equation like (8.8) is to know that it has been derived from an equation like (8.6), rather than being the original structural equation? A clue that suggests an underlying distributed lag model is responsible for Eq. (8.8) is the relationship between the serial correlation of errors in Eq. (8.8), and the coefficient of the lagged endogenous variable. A relationship of this kind would not be

---

[6] To derive the expression for the first-order serial correlation, recall that the numerator of the correlation between two variables is their covariance. In this case that will be the covariance of $u_t - qu_{t-1}$ with $u_{t-1} - qu_{t-2}$. Since the u_t are mutually uncorrelated, the only contributing term will be $-q\Sigma u_{t-1}$. The denominator of the correlation will be the variance of $u_t - qu_{t-1}$. Since $\Sigma u_t \simeq \Sigma u_{t-1}$, the variance will be $(1 + q^2)\Sigma u_{t-1}^2$.

expected when an equation of the form (8.8) represents the causal process itself.

The approach taken here can readily be modified to take account of deviations from a geometric distribution. In assuming that the influence of X on Y has the form of Eq. (8.6), we are assuming that a sudden change in X is felt immediately by Y, and then begins to die out immediately. If the effect of X dies out only after some time has elapsed, we would change the right-hand member of Eq. (8.6) to $bx_t + c(x_{t-1} + qx_{t-2} + q^2x_{t-3} + \ldots)$. This delays the onset of geometric decay by one lag. Equation (8.8) acquires a term in x_{t-1} when this is done.

At times, a causal effect is not felt with full force immediately, but takes some time to build up before it begins to decay. Such an effect can be modeled by adding a lagged endogenous variable to the original structural equation; the effect of doing this is to add a higher order autoregressive term to Eq. (8.8). Thus considerable flexibility in accomodating a variety of distributional schemes is possible.[7]

Correlations among Errors

We now examine the effect of serial correlation among error terms on our previous results. Two sets of assumptions concerning these correlations will be explored. In the first, we assume that only first-order serial correlation of errors is present, though the results will also hold when serial correlation of higher order is present, provided the order is finite. In the second, we assume that serial correlations of all orders are present and equal in size.

We show in the Appendix to this chapter that the general shape of the correlogram is unaffected by first-order serial correlation. Although serially correlated errors will change the numerical value of correlations, these changes are not easily detected from the general appearance of the correlogram. In the second case (that where serial correlations of all orders are present) the asymptotic values of the correlogram are changed. Instead of approaching zero asymptotically, the auto- and cross-correlations approach a nonzero constant at large n. When correlograms display this behavior, serial correlation of the second kind is present, and should be incorporated into one's model.

It is worth spending a moment to ask the question of what processes could produce errors with serial correlations that are equal at all orders. Suppose Z to be a variable that influences Y, but through misspecification

---

[7] For a fuller discussion see Rao and Miller (1971, pp. 161–173).

it has been omitted from the model, so that the error term consists of a term in Z and a random, independently distributed error term ξ that is uncorrelated with Z:

$$\epsilon_t = \beta Z_t + \xi_t.$$

We will then have

$$\text{cov}(\epsilon_t, \epsilon_{t-s}) = \beta^2 \text{cov}(Z_t, Z_{t-s}).$$

This expression will be equal to β^2, and thus will be independent of t, under three distinct circumstances:

1. Z is constant over time. This is equivalent to each individual in the sample having an individual and unchanging contribution to the criterion, in the amount βZ_i. In the study of biological phenomena, this might be a genetic endowment; in the social world, it might represent the lasting effect of a unique historical legacy.
2. Z_t increases or decreases linearly:

$$Z_t = \gamma_0 + \gamma_1 t.$$

   This is a trend effect.
3. Z_t is periodic, with a period given by T/n, where T is the interval between waves, and n is a positive integer. This sort of dependence would be found wherever strong daily, weekly, or seasonal variation is found.

Strict periodicty implies that a variable will have the same values at time $t + T$, $t + 2T$, $t + 3T$, etc., as at time t. This in turn implies that the correlations between one variable (whether periodic or not) and a periodic variable will be the same when the time lag separating the two variables is $t, t + T, t + 2T$, etc. For this reason, serially correlated errors having the form under consideration will lead to autocorrelograms and cross-correlograms that approach a constant at large lags, instead of approaching zero.

Trend effects and periodicity are expected only when working with unstandardized variables. If both the dependent variable and one or more independent variables introduced explicitly into the equation have a trend or periodicity, no problem is created. It is, rather, when an omitted explanatory variable has a trend or is periodic that we can expect error terms to have these features. Such effects can be detected by visual inspection of the error terms. When trends of periodicity appear in the error terms, it would be appropriate either to introduce these effects explicitly into the model, or to take care of them by introducing serial correlation of errors explicitly.

Overview

The analytic investigation reported here shows that empirical analysis of cross-correlograms can constrain the possibilities the analyst needs to consider when studying the sensitivity of model parameters to different specifications of causal lag. However, this type of empirical diagnosis is incapable of pinpointing precisely the lag structure in one's data. At best, in the use of partial cross-correlograms, it is possible to bound the cross-lags at an upper limit.

In the next chapter we demonstrate how this information about bounds can be used to study the sensitivity of parameter estimates to variations of the lag structure within the range of permissible values defined by the partial correlogram analysis.

Appendix: The Analytical Properties of the Correlogram

In this appendix we derive algebraic formulas that express the correlations r_n, r_n^*, s_n, and t_n in terms of the model parameters a, b, c, and d (as defined in Eqs. [8.1] and [8.2]). To do this we work with the recurrence relations (8.4a)–(8.4d).

These equations are nontrivial because they are coupled. The equation for r_n depends on s_{n-1} as well as on r_{n-1}, while the equation for s_n depends on r_{n-1} as well as on s_{n-1}. We outline two different methods for handling this difficulty.

In the first, we start by decoupling the equations. This is done by eliminating s_{n-1} from Eqs. (8.4a) and (8.4b), thus providing an expression for cs_n in terms of r_n and r_{n-1}:

$$cs_n = ar_n + (ab - cd)r_{n-1}.$$

Lagging this expression by one time unit, we obtain an equation for cs_{n-1}. When this expression is inserted in Eq. (8.4b), we find

$$r_n - (a + b)r_{n-1} + (ab - cd)r_{n-2} = 0. \tag{A8.1}$$

It follows from symmetry considerations that r_n^*, s_n, and t_n all obey this same equation (with the same coefficients). This greatly simplifies our work.

Equation (A8.1) is a linear, second-order, homogeneous difference equation with constant coefficients. Its most general solution (Goldberg, 1958, pp. 134–143) is

$$r_n = C_1 m_1^n + C_2 m_2^n \tag{A8.2}$$

where C_1 and C_2 are arbitrary constants and m_1 and m_2 are the two solutions to the quadratic equation

$$m^2 - (a + b)m + (ab - cd) = 0.$$

These solutions are

$$m_1 = \{(a + b) + \sqrt{(a - b)^2 + 4cd}\}/2, \qquad (A8.3a)$$

$$m_2 = \{(a + b) - \sqrt{(a - b)^2 + 4cd}\}/2. \qquad (A8.3b)$$

The behavior of the solutions as n varies depends on the character of the roots m_1 and m_2. Three mutually exclusive cases exhaust the possibilities:

1. m_1 and m_2 are real and unequal. This will occur whenever the quantity $(a - b)^2 + 4cd$ is positive. We note that if m_1 or m_2 were to exceed 1 in magnitude, r_n would grow without limit as n increases. Since a correlation cannot exceed 1 in magnitude, we are restricted to the case where both roots are less than or equal to 1. This will be the case provided $cd < (1 - a)(1 - b)$.

Ordinarily, the quantities a and b will be positive. Under this circumstance, m_1 and m_2 will be positive provided that $ab > cd$. Usually, stability coefficients will be larger in magnitude than cross-coefficients, and so this inequality will typically hold. As long as it does, all powers of m_1 and m_2 will be of the same sign. It can be shown that under this circumstance r_n has only one maximum (or minimum) and can cross the origin at most once. At large n, r_n will be dominated by the larger root, m_1. Consequently, r_n will decline to zero geometrically at large n.

2. $m_1 = m_2$, both roots being real. This will occur only when the quantity $(a - b)^2 + 4cd = 0$. In this case, the solution becomes

$$r_n = (C_1 + C_2 n)m^n.$$

The solution will be positive for all n if $(a + b)$ is positive, and will decline geometrically to zero at large n.

3. m_1 and m_2 are complex. This will occur if $(a - b)^2 + 4cd$ is negative. From the expressions for the roots in Eqs. (A8.3a) and (A8.3b), we see that m_1 and m_2 are complex conjugates of one another. We can, therefore, write

$$m_1 = m_R + im_I, \quad m_2 = m_R - im_I,$$

where

$$m_R = (a + b)/2, \quad m_I = \sqrt{-(a - b)^2 - 4cd}, \quad \text{and} \quad i = \sqrt{-1}.$$

By making use of the properties of imaginary numbers, we can rewrite the solution given in Eq. (A8.2) in the form

$$r_n = \rho^n \{ C_1' \cos(n\theta) + C_2' \sin(n\theta) \}$$

where

$$\cos \theta = m_R / \sqrt{m_R^2 + m_I^2}, \quad \sin \theta = m_I / \sqrt{m_R^2 + m_I^2},$$
$$\rho = \sqrt{m_R^2 + m_I^2} = \sqrt{ab - cd}, \quad C_1' = C_1 + C_2,$$

and

$$C_2' = i(C_1 - C_2).$$

Here the correlations oscillate in sign. When $\rho = 1$, the solutions are undamped (the magnitude of the correlations does not decline as n increases), while with $\rho < 1$ the solutions are damped and approach zero asymptotically as n becomes very large.

The constants C_1 and C_2 (or equivalently, C_1' and C_2') are chosen so that the solution will reproduce the correlations r_0 and r_1. Assuming we are not dealing with case 2, where both roots are real and equal, we must choose those values of C_1 and C_2 that satisfy the simultaneous linear equations

$$C_1 + C_2 = r_0 = \frac{ac + bd}{1 - ab - cd},$$

$$C_1 m_1 + C_2 m_2 = r_1 = \frac{c + d(b^2 - c^2)}{1 - ab - cd}.$$

Solving for C_1 and C_2, and plugging in the expressions (A8.3a)–(A8.3b) for m_1 and m_2, we find

$$C_1 = \frac{-(ac + bd)[a + b - \sqrt{(a - b)^2 + 4cd}] + 2[c + d(b^2 - c^2)]}{2(1 - ab - cd)\sqrt{(a - b)^2 + 4cd}},$$

$$C_2 = \frac{(ac + bd)[a + b + \sqrt{(a - b)^2 + 4cd}] - 2[c + d(b^2 - c^2)]}{2(1 - ab - cd)\sqrt{(a - b)^2 + 4cd}}.$$

Similarly, the coefficients for s_n are obtained by solving the equations

$$C_1^* + C_2^* = s_0 = 1,$$

$$C_1^* m_1 + C_2^* m_2 = s_1 = \frac{a + b(d^2 - a^2)}{1 - ab - cd};$$

in this case, the solutions are

$$C_1^* = \frac{1}{2} + \left[\frac{a + b(d^2 - a^2)}{1 - ab - cd} - \frac{(a + b)}{2} \right] / \sqrt{(a - b)^2 + 4cd},$$

$$C_2^* = \frac{1}{2} - \left[\frac{a + b(d^2 - a^2)}{1 - ab - cd} - \frac{(a + b)}{2} \right] / \sqrt{(a - b)^2 + 4cd}.$$

In an alternative approach, we define the two-dimensional vector

$$V_n = \begin{pmatrix} s_n \\ r_n \end{pmatrix}.$$

We can then write Eqs. (8.4a) and (8.4b) as $V_n = MV_{n-1}$, where

$$M = \begin{pmatrix} a & d \\ c & b \end{pmatrix}.$$

Iterating the matrix equation, we have $V_n = M^n V_0$. Since V_0 is already known, we can find s_n and r_n for any integral value of n by finding the nth power of M, and post-multiplying by V_0. The only question remaining is how one finds the nth power of a 2×2 matrix.

To find the nth power of M, we make use of the fact that an arbitrary square matrix M can be expressed as the matrix product $S\Lambda S^{-1}$, where Λ is the diagonal matrix

$$\Lambda = \begin{pmatrix} \lambda_1 & 0 \\ 0 & \lambda_2 \end{pmatrix}.$$

To find λ_1 and λ_2 and the matrix S, we multiply the matrix equation $M = S\Lambda S^{-1}$ on the right by S, to obtain $MS = S\Lambda$, from which $(M - \Lambda)S = 0$. This equation holds only if the determinant of $M - \Lambda$ vanishes. We thus set

$$\begin{vmatrix} a - \lambda & d \\ c & b - \lambda \end{vmatrix} = 0.$$

Computing the determinant, we obtain the quadratic equation

$$\lambda^2 - (a + b)\lambda + (ab - cd) = 0.$$

This is the same equation previously derived in this appendix for m, and its solutions λ_1 and λ_2 coincide with the roots m_1 and m_2 in Eqs. (A8.3a)–(A8.3b). The elements of the matrix S are determined by inserting the expressions for λ_1 and λ_2 just obtained in the equation $MS = \Lambda S$. The solution, which is defined only up to a multiplicative constant, is

$$S = \begin{pmatrix} 1 & 1 \\ \dfrac{-(a - b) + \sqrt{(a - b)^2 + 4cd}}{2d} & \dfrac{-(a - b) - \sqrt{(a - b)^2 + 4cd}}{2d} \end{pmatrix}.$$

Next, we note that $M^2 = S\Lambda S^{-1}S\Lambda S^{-1} = S\Lambda^2 S^{-1}$. Similarly, $M^n = S\Lambda^n S^{-1}$. Our problem has been reduced to finding the nth power of Λ. But since Λ is diagonal, this is much easier than finding the nth power of M, which is the most general 2×2 matrix. We simply have

$$\Lambda^n = \begin{pmatrix} \lambda_1^n & 0 \\ 0 & \lambda_2^n \end{pmatrix}.$$

From here it is a simple matter of matrix multiplication to compute $V_n = S\Lambda^n S^{-1} V_0$.

Extensions

Distributed Lags

Extension of the above derivations to the case where simultaneous as well as lagged cross-effects are present is direct. If the structural equations are

$$X_t = aX_{t-1} + dY_{t-1} + eY_t + \epsilon_t$$
$$Y_t = bY_{t-1} + cX_{t-1} + fX_t + \nu_t,$$

they can be manipulated to obtain the reduced equations, in which all predictor variables are lagged:

$$Y_t = \{(c + af)X_{t-1} + (b + df)Y_{t-1} + (\nu_t + f\epsilon_t)\}/(1 - ef)$$

and

$$X_t = \{(a + cd)X_{t-1} + (d + be)Y_{t-1} + (\epsilon_t + e\nu_t)\}/(1 - ef).$$

From here one can substitute coefficients in the results already derived, e.g., $a \to a + cd$, etc.

If distributed lags are present in one's model (e.g., effects involving X_{t-2}, Y_{t-2}, etc., as well as X_{t-1} and Y_{t-1}), the approach taken here can still be followed, but one obtains a difference equation of higher order than 2, or a more complicated matrix equation. In principle, higher order difference equations can be solved in the same way as the second-order equation, but one finds an nth order polynomial for the roots m, instead of a quadratic. In general, there will be n roots, and except in special cases, only numerical solutions will be possible. However, if complex roots occur they will still occur in complex conjugate pairs, and so oscillatory solutions are still possible.

Serially Correlated Errors

We now introduce serial correlation among error terms. To avoid irrelevant complexity, we suppose causal influences to be unidirectional, and take $d = 0$ in Eq. (8.1). Next, we will comment on how our findings change when this restriction is dropped.

First we consider first-order serial correlation of errors:

$$\nu_t = \rho\nu_{t-1} + \phi_t,$$ (A8.4)

with ϕ_t being independently distributed and uncorrelated with ν_{t-1}, X_{t-1}, and Y_{t-1}. Error terms are assumed to be cross-sectionally uncorrelated, and the ϵ_t are assumed to be serially uncorrelated (the parameter a would be under-identified if the ϵ_t were to be serially correlated).

The first step is to find an expression for $\text{cov}(Y_t, \nu_{t-n})$. From Eq. (8.2),

$$\text{cov}(Y_t, \nu_t) = b \, \text{cov}(Y_{t-1}, \nu_t) + \nu_t^2 = b\rho \, \text{cov}(Y_{t-1}, \nu_{t-1}) + \nu_t^2.$$

Making use of the stationarity condition, we have

$$\text{cov}(Y_t, \nu_t) = \nu_t^2/(1 - b\rho).$$

It then follows from Eq. (A8.4) that

$$\text{cov}(Y_t, \nu_{t-n}) = \rho^n \nu_t^2/(1 - b\rho).$$

Instead of Eq. (8.4d), we now have

$$t_n = bt_{n-1} + cr_{n-1}^* + \rho^n \nu_t^2/(1 - b\rho);$$

the other normal equations, (8.4a)–(8.4c), remain unchanged.

The second of the two derivations carried out for the case where serially uncorrelated errors were assumed, can now be extended to the present case. We define a three-dimensional vector

$$V_n = \begin{pmatrix} t_n \\ r_n^* \\ \rho_n \end{pmatrix}$$

where

$$\rho_n = \rho^n \nu_t^2/(1 - b\rho).$$

The normal equations are then expressed in matrix form as in the preceding equations, $V_n = MV_{n-1}$, but now

$$M = \begin{pmatrix} b & c & \rho \\ 0 & a & 0 \\ 0 & 0 & \rho \end{pmatrix} = \rho \begin{pmatrix} a/\rho & c/\rho & 1 \\ 0 & a/\rho & 0 \\ 0 & 0 & 1 \end{pmatrix}.$$

Iterating the matrix equation, we have $V_n = M^n V_0$, where

$$V_0 = \begin{pmatrix} 1 \\ ac/(1 - ab) \\ \nu_t^2/(1 - b\rho) \end{pmatrix}.$$

We now make use of some results from the theory of finite Markov chains (Kemeny and Snell, 1960; Fararo, 1973, pp. 290–295; Greenberg, 1979, pp. 242–244). A square matrix that has the form

$$M = \begin{pmatrix} R & Q \\ 0 & I \end{pmatrix}$$

has, as its nth power,

$$M^n = \begin{pmatrix} R^n & (I - R)^{-1}Q \\ 0 & I \end{pmatrix}.$$

In the case at hand,

$$R = \begin{pmatrix} b/\rho & c/\rho \\ 0 & a/\rho \end{pmatrix} \quad \text{and} \quad Q = \begin{pmatrix} 1 \\ 0 \end{pmatrix}.$$

By direct computation,

$$(I - R)^{-1}Q = \begin{pmatrix} 1 \\ 0 \end{pmatrix}/(1 - b/\rho).$$

It follows that t_n will contain a term in $\rho^n v_t^2/(1 - b/\rho)(1 - b\rho)$. As n increases, this term declines geometrically. Thus the asymptotic behavior of t_n at large n is unaffected by first-order serial correlation of errors. It is obvious that this conclusion is equally valid if serial correlations of any finite order are present among the disturbances.

The second error structure we consider is defined by $\text{cov}(v_t, v_{t-s}) = \mu$, where μ is a constant independent of s. This structure differs from the one just considered in that it implies serial correlation of *infinite* order among the error terms.

Now we have

$$\text{cov}(v_t, Y_{t-n}) = \begin{pmatrix} v_t^2/(1 - \mu b) \\ \mu v_t^2/(1 - \mu b) \end{pmatrix} \quad \begin{array}{l} \text{when } n = 0, \\ \text{when } n \neq 0. \end{array}$$

Equation (8.4d) now takes the form

$$t_n = bt_{n-1} + cr_{n-1}^* + \mu v_t^2/(1 - \mu b), \quad n \geq 1.$$

Repeating the previous procedure, we find that serial correlation supplies a term $\mu v_t^2/(1 - b)(1 - \mu\beta)$ to t_n. This term does not vanish as n increases without limit.

In the more general case where $d \neq 0$, these results still hold. However, if there is cross-sectional correlation between the errors of X_t and the errors of Y_t, this will "transmit" serial correlation in one set of errors to the cross-correlogram in which X is lagged behind Y, as well as to the autocorrelogram for the X_t.

9

Sensitivity Analysis

Many of the issues we have considered in earlier chapters center on problems of identification: of determining the assumptions the researcher has to make to find unique estimates of the parameters in a particular model or part of a model. In Chapter 3, for instance, we showed that cross-sectional and two-wave panel models usually must rely on assumptions that certain regression coefficients have fixed values. We also showed that at times multiwave panel models could be identified with far weaker assumptions, namely, that certain coefficients remain constant over time or have some other given time dependence.

Occasionally, though, these assumptions cannot be made. In a cross-sectional or two-wave panel study, there may be no plausible grounds for fixing any of the regression coefficients at any particular value. In a multiwave study, it may not be reasonable even to assume that certain effects are constant. The regularity in the change process that such an assumption entails is not always warranted. For example, Kenny and Harackiewicz (1979) suggest that the causal relationships that govern a child's development change as the child grows. In a period of revolutionary upheaval just after a new nation is established, the social processes that link key political and economic variables may not be stable enough to warrant the assumption that coefficients do not change.

When neither constancy constraints nor assumptions about the values of individual regression coefficients can be justified, the model is underidentified,[1] and so the parameters of the model cannot be uniquely esti-

---

[1] Tests to determine identifiability are discussed in Namboodiri, Carter, and Blalock (1975, p. 501), Asher (1976, pp. 49–61), and Fisher (1966), among others.

mated from the observed data.[2] Two practices are common at this stage: to give up, or to impose identifying restrictions arbitrarily, without real theoretical justification, and hope for the best. There is another option, though: to investigate the extent to which information about the range of values various parameters in the model can assume constrains other parameters. Analyses of this kind are called *sensitivity analyses* because they determine how sensitive estimates of some parameters are to assumptions made about other parameters in the model.

The hope in such analyses is that estimates of parameters about which the analyst is uncertain will prove to depend only weakly on the assumptions made about other parameters. In that case it may be possible to confine the parameters of interest within a narrow range even if a unique estimate cannot be obtained.

Sensitivity analyses can be carried out in several ways. One is to re-estimate the parameters of interest using different assumptions than those employed in the original analysis. As we noted in Chapter 3, Kohn and Schooler conducted a sensitivity analysis of this sort by constraining the parameters of different background variables in their model a variety of ways, thus arriving at several different estimates of the same parameters. As it happened, these estimates were all very similar to each other, thus increasing the researchers' confidence in the robustness of their findings.

When the additional over-identifying information required to re-estimate models is not available, it will almost always be possible to focus on a range of plausible values for a parameter assumed in the original model to have a given value. When this can be done, it is possible to re-estimate the model by varying the assumed value of this critical parameter within the range considered plausible. For example, Duncan (1969b) estimated models in a study of motivation and work achievement by adopting a series of fixed values for those parameters whose values could not be specified precisely on theoretical grounds. We adopted this approach in Chapter 6 when we reanalyzed Eaton's data on life events and psychological distress.

A third approach, suggested by Land and Felson (1978), consists of imposing inequalities on some coefficients and then deriving inequalities for other coefficients through the algebraic manipulation of path equations. This approach is expected to be useful when one does not know the value of a parameter a priori, but is confident of its sign or that it falls within a given range. Land and Felson point out that this approach can be generalized through the use of linear and nonlinear programming techniques to handle constraints that take the form of equalities or inequalities

[2] It can happen that a model as a whole is under-identified even though *some* of its parameters are identified or even over-identified.

for linear combinations of variables.[3] Using this procedure the researcher can bracket unindentified parameters between lower and upper limits.

None of these approaches to sensitivity analysis is unique to panel data, since none of them makes use of time in a fundamental way. So we do not develop general guidelines for the use of these approaches here. We have introduced the topic, though, because sensitivity analysis is a particularly useful strategy for two general problems faced by the researcher working with panel data.

The first problem is that of estimating cross-lagged and cross-contemporaneous effects of X_t and Y_t in a two-wave, two-variable model. The solutions to this problem offered in Chapter 3 required conditions that will not always obtain. It is consequently useful to investigate what can be learned about lagged and instantaneous cross-effects in a less ideal situation.

The second problem we consider is that of a misspecified causal lag in a structural model. Making use of the results reported in Chapter 8, we show how models can be reestimated across a range of plausible lags. The technique we present can also be useful when we know the true causal lag but have data collected at time intervals that do not correspond to the true lag.

We restrict our discussion of sensitivity analysis to strategies that are particularly appropriate to these problems.

The Two-Wave, Two-Variable Model Revisited

We noted in Chapter 3 that the most general two-wave, two-variable model (see Figure 3.1) is under-identified (Duncan, 1969a), since the six independent correlations among the four variables X_1, Y_1, X_2, and Y_2 are insufficient to estimate the eight parameters that characterize the model. Yet one might wonder whether a sensitivity analysis would permit some information about the parameters to be extracted on the basis of two waves of data, even if unique estimates cannot be obtained.

Let us suppose that X and Y have both lagged and instantaneous reciprocal effects on one another, and that time 2 error terms may be correlated. We can write the structural equation for Y_2, assuming for the sake of convenience that variables have been standardized, as follows:

$$Y_2 = aY_1 + bX_1 + b\lambda X_2 + v \qquad (9.1)$$

where the instantaneous effect of X on Y has been expressed as a multiple

[3] For an introduction to linear and nonlinear programming, see Chiang (1974, pp. 621–741).

of its lagged effect. When $\lambda = 0$, the instantaneous effect of X_2 on Y_2 vanishes; when $\lambda = \pm\infty$, the lagged effect of X on Y is vanishingly small.

The presence of instantaneous reciprocal effects of X and Y on one another precludes the use of X_2 to help identify Eq. (9.1), but X_1 and Y_1 can be used as instruments, so that the parameters a and b can be expressed in terms of the observed correlations and the parameter λ. On the assumption that cross-effects are of greatest interest, we can forget about the estimate of a for the moment, and concentrate on the estimate of b, which is

$$\hat{b} = \frac{(r_{X_1Y_2} - r_{X_1Y_1}r_{Y_1Y_2})}{(1 - r_{X_1Y_1}^2) + \lambda(r_{X_1X_2} - r_{Y_1X_2}r_{X_1Y_1})}. \tag{9.2}$$

This equation shows that the estimate of b is a function of the values of observed correlations and the value of λ. If the precise value of λ is known to the investigator, this value can be inserted into Eq. (8.3) to yield an unbiased estimate of b. Commonly, of course, the investigator will not know the precise value of λ. But even partial information can be useful in estimating the range of values b might take on.

One approach that can be useful is to estimate b using a series of different values of λ; this is an application of Duncan's approach to sensitivity analysis described above. When these values span the range of plausible values of λ, the range of estimated values for b will represent the upper and lower bounds for this parameter. For instance, we might be willing to assume that the lagged and contemporaneous effects of X_1 and X_2 on Y_2 are of the same sign, but that the lagged effect is larger than the contemporaneous effect. In this case, the constraint imposed on λ is $0 \leq \lambda < 1$, and we can estimate a series of values for $\hat{b}$ by entering a series of discrete values lying between 0 and 1 into Eq. (9.2).

As we indicated in Chapter 3, most researchers working with two-wave panel data make the much more restrictive assumption that contemporaneous cross-effects are entirely absent; in other words, that $\lambda = 0$. When this assumption is in error it can happen that not even the sign of b will be estimated correctly by Eq. (9.2). We can see why this is so by considering this equation in more detail.

Since the numerator in Eq. (9.2) does not depend on λ, and the first term in the denominator is positive, the sign of $\hat{b}$ will depend on the sign of the quantity $\lambda(r_{X_1X_2} - r_{Y_1X_2}r_{X_1Y_1})$ and its magnitude relative to the quantity $(1 - r_{X_1Y_1}^2)$. Thus, if we assume incorrectly that $\lambda = 0$, Eq. (9.2) will yield the correct sign for b provided that

(a) $\lambda \geq 0$ and $r_{X_1X_2} \geq r_{Y_1X_2}r_{X_1Y_1}$

or

(b) $\lambda \leq 0$ and $r_{X_1X_2} \leq r_{Y_1X_2}r_{X_1Y_1}$.

On the other hand, if

$$\text{(c)} \quad \lambda > 0 \quad \text{and} \quad r_{X_1 X_2} < r_{Y_1 X_2} r_{X_1 Y_2}$$

or

$$\text{(d)} \quad \lambda < 0 \quad \text{and} \quad r_{X_1 X_2} > r_{Y_1 X_2} r_{X_1 Y_1},$$

Eq. (9.2) may or may not yield the correct sign of b, depending on the magnitude of λ relative to the quantity

$$\lambda' = (1 - r_{X_1 Y_1}^2)/(r_{X_1 X_2} - r_{X_2 Y_1} r_{X_1 Y_1}). \tag{9.3}$$

If the magnitude of this quantity is larger than the magnitude of λ, the estimate of b made on the basis of Eq. (9.2) (with λ set equal to zero) will yield the correct sign of b; otherwise not.

In most of the panel studies we have examined, the stability of each variable has been reasonably high, so that the inequality

$$r_{X_1 X_2} \geq r_{Y_1 X_2} r_{X_1 Y_1} \tag{9.4}$$

held. We would generally expect this to be the case. Where Eq. (9.4) is valid, the relevant possibilities are (a) and (d). In the former case, the two-wave model yields the correct sign of the cross-effect; in the latter case, one can be confident that this is so only if the magnitude of expression (9.3) is substantially larger than any plausible value of λ. In empirical applications, this expression need not be very large, in which case a degree of knowledge about the relative magnitude of lagged and instantaneous effects ordinarily unavailable to the investigator would have to be available to justify estimating the sign of b on the basis of Eq. (9.2). It may be recalled that no such information was required when three waves of data were available.

Where a priori information about λ is unavailable to the investigator, it may be possible to restrict its value somewhat by confining the autoregressive coefficient a in Eq. (9.1) to the range between 0 and 1; we would normally expect this restriction to be valid. Since the estimator for a is

$$\hat{a} = \frac{\{(r_{Y_1 Y_2} - r_{X_1 Y_2} r_{X_1 Y_1}) + \lambda(r_{X_1 X_2} r_{Y_1 Y_2} - r_{X_1 Y_2} r_{X_2 Y_1})\}}{\{(1 - r_{X_1 Y_1}^2) + \lambda(r_{X_1 X_2} - r_{X_1 Y_1} r_{X_2 Y_1})\}}, \tag{9.5}$$

the requirement $0 \leq a \leq 1$ leads to two inequalities for λ. In some cases we may be able to constrain the permissible values of λ on the basis of these inequalities sufficiently to estimate whether or not the sign of b is correctly estimated in Eq. (9.2) under the assumption that $\lambda = 0$.

To illustrate the application of these procedures, we return once more (for the last time) to the Kohn and Schooler study of work complexity and intellectual flexibility, and to the Greenberg, Kessler and Logan study of crime rates and arrest rates, both discussed previously in Chapter 3.

Work Complexity and Intellectual Flexibility

We begin with the data of Kohn and Schooler, but rather than assume as they did that both lagged and contemporaneous cross effects might exist, we make the more restrictive assumption that only lagged effects exist. The correlations among intellectual flexibility (IF) and substantive complexity of work (SC) measured in 1964 and 1974 are given in Table 9.1. The path equation corresponding to the assumption that contemporaneous cross effects are zero are then

$$SC_2 = .41SC_1 + .44IF_1,$$
$$IF_2 = .14SC_1 + .77IF_1.$$

The first step in our analysis of the sensitivity of the cross coefficients is to see whether inequality (9.4) holds, and we find that it is indeed valid, since the left-hand member is .74 while the right-hand member is 0.56 (using the correlations in Table 8.1 and focussing only on the effects of work complexity on intellectual flexibility). This restricts us to cases (a) and (d) above, so that our evaluation of the correctness of the sign estimated for b leads one to an assessment of the probable sign of λ. Since it is difficult to think of a reason that the lagged and instantaneous effects of work complexity on intellectual flexibility would have opposite signs, it seems reasonable to reject the possibility that λ is negative. As a result, the only case that is plausible is (a). In this case, the estimate of b is unbiased in sign in the two-wave model, a gratifying finding.

Table 9.1
Correlation Matrix for Work Complexity (SC)
and Intellectual Flexibility (IF) at Two Times[a]

	SC_1	SC_2	IF_1	IF_2
SC_1				
SC_2	.74			
IF_1	.75	.75		
IF_2	.71	.67	.87	

[a] These correlations have been computed from the estimates Kohn and Schooler (1978) obtained for the model we discussed in Chapter 3.

Crime Rates and Arrest Rates

To show that it will not always be possible to constrain the sign of b so easily, we apply the sensitivity analysis procedure to the first two waves of the four-wave model of crime rates and arrest rates studied by Greenberg, Kessler, and Logan (1979). Identifying X with the arrest rate and Y with the total offense rate, we ask ourselves whether a two-wave cross-lagged model yields the correct sign for the lagged effect of arrests on crime.

We begin by inserting the appropriate correlations from Table 9.2 into inequality (9.4), and we find that the inequality holds. Thus we again restrict our attention to cases (a) and (d).

In this case, we may be less confident that λ is nonnegative. We may think, for example, that arrests have a short-run tendency to reduce crime by taking violators out of circulation; but that the long-run effect of arrests is to stigmatize those arrested, increasing the crime rate. Of course, we may not be certain of this; the possibility that arrests have a long-run tendency to reduce crime rates through their deterrent effects is something we would not want to exclude a priori. Thus, both cases (a) and (d), corresponding to negative or nonnegative values of λ, must be considered.

Since theoretical considerations fail to restrict λ, we restrict the parameter a to lie between 0 and 1 (making use of Eq. [9.5] for ($\hat{a}$)) to obtain restrictions on λ. These restrictions yield two inequalities for λ. These inequalities restrict λ to be less than -1.77.

Since these inequalities permit both negative and positive values of λ, neither case (a) nor case (b) can be excluded. In the former case, of course, b is estimated without bias on the basis of the assumption that λ is zero; in the latter case, however, expression (9.3) must be examined to

Table 9.2
Correlation Matrix for Total Reported Index Offense Rate (Y) and Arrest Clearance Rate (X) at Times 1 and 2[a]

	X_1	X_2	Y_1	Y_2
X_1		.637	$-.346$	$-.355$
X_2			$-.286$	.928
Y_1				$-.355$
Y_2				

[a] For a description of the data from which these correlations were computed, see Greenberg, Kessler, and Logan (1979).

determine whether the sign of b can be estimated without bias from Eq. (9.2) assuming that $\lambda = 0$. Inserting the observed correlations in Eq. (9.3), we find $\lambda' = 1.64$. The sign of the estimator of b will be unbiased only if the magnitude of λ' is larger than the magnitude of λ. As the observed magnitude of λ is[4] -2.15 ($= -.260/.121$), we have case (d), and since the required inequality for λ does not hold, the sign of b obtained on the basis of the two-wave model would be in error.

This finding substantiates the argument made in Chapter 3 that two-wave cross-lagged models can be seriously misleading when true causal coefficients are both instantaneous and lagged. This is especially true, as we see here, when the lagged and instantaneous effects are of opposite sign.

Incorrect Causal Lag

In analyzing a given set of panel data, the researcher may discover on the basis of cross-correlogram analysis (see Chapter 8) that the causal lag appropriate for the research problem can be constrained only within a relatively wide range. If the causal model estimated embodies all the correct causal relationships, but is estimated with an incorrect causal lag, parameter estimates can be biased. The question then arises: can the sensitivity of parameters to a misspecification of the causal lag be investigated? Note that this problem differs from the one encountered in initial model specification. There one seeks a method for inferring the lag from the observed correlations. Now we suppose that a range of possible lags is theoretically plausible, and we want to investigate the variation in values of the parameter estimates corresponding to the different possibilities in this range.

If the true lag lies within a range that is *longer* than the interval between observations, but shorter than the total time interval covered in the multiwave panel, no special difficulty arises. Thus if one believes that the true lag lies between 3 and 6 months, and observations have been collected monthly for 6 months, one simply estimates the model across the range of potential intervals. If the true lag lies beyond the time interval of the study, then the situation is hopeless. When the range of possible lags lies within the time period of the study, but is *shorter* than the interval between observations, the problem is difficult but solvable.

To make the discussion concrete, suppose the interval between obser-

---

[4] For purposes of this discussion, we neglect the issue of sampling error in the estimation of λ. The estimate itself is unbiased, and is computed from the figures in Table 3.2.

vations is an integral multiple of the lags we want to investigate.[5] Thus, we might want to investigate how our parameter estimates change if we assume that the true causal lag is 1, 2, 3, and 4 months, but have observations collected only annually. Our task is to estimate the model parameters for each of these possible lags.

To begin, we use the structural equations to generate expressions for the autocorrelograms and cross-correlograms. This is just the sort of thing we did in the Appendix to Chapter 8 for the cross-lagged bivariate panel model. Once these expressions have been derived, we choose the autocorrelations with lags corresponding to the intervals between observations, and discard the rest. For instance, if we are studying the solution for the 1-month lag, we choose the autocorrelations and cross-correlations for lag 12, since the observed 1-year lag is actually twelve 1-month lags beyond the causal interval being considered. We then set the expressions for the correlations retained in this procedure equal to the observed correlations. This yields a set of nonlinear equations for the model parameters. As long as the model is identified in the observed scores, ignoring the temporal misspecification, it will be possible to solve these equations for the parameters.

If the model is at all complex, or if many waves of predicted correlations must be computed, the labor this procedure requires may be considerable. However, estimates can be obtained using LISREL; here one simply estimates a model in which the variables at times when no observations are collected are treated as unmeasured, latent variables. Equality constraints on the coefficients of such models can be used to identify such models, just as when all variables are observed.

This procedure is repeated for each value of the lag considered. At the end of the entire series of calculations it is possible to compare the parameter estimates corresponding to each lag to determine the sensitivity of the estimates to temporal misspecification.

This approach can also be used when we have reason to believe that the true causal lag does not correspond to the time interval between observations. We illustrate the procedure by applying it to the model of Wheaton et al. (1977) analyzed in Chapter 2. In this model, the interval between observations at time 1 and time 2 was 1 year, while the interval between observations at times 2 and 3 was 4 years. Suppose we had reasons for thinking that the actual causal lag was in fact 1 year. We would then have no qualms about interpreting the estimates of parameters that link times 1 and 2. But we would be uncertain as to the interpretation of the parameters linking times 2 and 3. Moreover, any comparison between the two

[5] We will comment later on what to do when this assumption cannot be made.

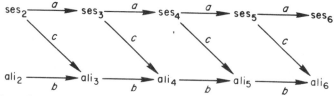

Figure 9.1. Panel model for SES and alienation. Variables with subscripts 2 and 6 correspond, respectively, to time 2 and time 3 observations in the original study of Wheaton *et al.* (1977); variables with subscripts 3, 4, and 5 are unobserved. To avoid confusion from the two sets of subscripts, lower case letters are used to denote socioeconomic status and alienation here, while upper case letters are used for the observed variables. Thus $ali_2 = A_2$, $ali_6 = A_3$, and similarly for ses.

sets of parameters would be complicated by the difference in intervals between observations. If we find more change over 4 years than over 1, is this because more time has elapsed, or because the causal processes themselves have changed?

Let us suppose we felt confident that the actual causal lag was in fact one year. We would then use the model displayed in Figure 9.1 to estimate the observed correlations linking observations at times 2 and 3.

Since causal influences in Figures 2.2 and 9.1 are assumed to be unidirectional and lagged only, we can use the expressions in Eqs. (8.5a)–(8.5d) to generate expressions for the correlations involving lags of four time units (4 years in this case). We identify ses with X and alienation with Y. Since Wheaton *et al.* assume socioeconomic status to remain the same over the duration of the study, we can set $a = 1$, and since four time units separate the second and third waves of observations, we set $n = 4$.

The expressions in Eqs. (8.5a)–(8.5d) simplify, and we are left with

$$r_4 = r_{ali_6,ses_2} = c/(1 - b) = r_{A_3,SES_2} = -.620$$
$$t_4 = r_{ali_6,ses_2} = b^4 - c^2(b^4 - 1)/(1 - b)^2 = r_{A_3,A_2} = .701.$$

This gives us two algebraic equations for the parameters c and b; they can be solved to yield[6] $b = .846$, $c = -.096$.

These values can be compared with those that link time 1 and time 2 observations (see Figure 2.3). The parameters for that part of the model are $b = .712$ and $c = -.183$. Comparing the parameters that link time 2 and time 3 with those that link time 2 with time 3, we see that alienation be-

---

[6] These parameter estimates meet the criterion given earlier in this chapter for displacement of the peak in the correlogram of cross-correlations. Thus there is no contradiction between our assumption that the correct causal lag is one time unit and the pattern of correlations in Table 2.1, and our assumption that the correct causal lag is one year, even though the correlation between SES and A_3 is larger in magnitude than the correlation between SES and A_2.

comes slightly more stable, and is influenced slightly less by socio-economic status as time goes on, but that these changes are quite small. Substantively, the two sets of parameters seem consistent with one another. A comparison of this kind was not possible using the methods employed by Wheaton *et al.*, or in the approach we utilized in reanalyzing their work in Chapter 2.

When the interval between observations is *not* an integral multiple of the lag in the theoretical model, there are two ways of proceeding. The first is to use differential equation methods; these are discussed in Chapter 11. The second method is to estimate two sets of models. Suppose the ratio of time between observed and true lag is Q. The procedure just described would be carried out under the assumption that this ratio is the largest integer smaller than Q, and repeated under the assumption that the ratio is the smallest integer larger than Q. Linear interpolation would then be used to obtain a single best set of parameter estimates from the two sets. Thus, if we believed the true causal lag to be 5 months, but observations had been collected annually, we would have $Q = 12/5 = 2.4$. We would then estimate our model assuming $Q = 2$, corresponding to a true lag of 6 months, and again assuming $Q = 3$, corresponding to a true lag of 4 months. To regain the estimate corresponding to the 5-month lag, we would average these two estimates.

This procedure can be used with any panel collected over a longer time interval than the researcher considers plausible for the true causal lag. It should not be concluded, though, that the optimal panel design is therefore to wait as long as possible before gathering the second wave of data. This is so for two reasons. First, the procedure is predicated on the assumption that the structural coefficients are stationary. While this will often be true, in practice, this assumption is increasingly suspect as the time interval approaches a length over which we would expect historical change (in the case of aggregate data) or life-cycle change (in the case of individuals) to take place.

Second, even when stationarity can safely be assumed, extension of the temporal duration of a panel wave reduces the sensitivity of parameter estimates. A two-wave study carried out over a time interval of 6 months can study the possibility that the true lag is 2 or 3 months more effectively than is possible with panel data collected with a time interval of 6 years between waves.

Together, these two reasons argue that the researcher designing a panel study should attempt to collect data as close to the presumed causal interval as possible. While this simple rule leaves many questions unanswered (for example: How do I know the true causal interval before collecting the data?), we postpone consideration of them until Chapter 12, where we discuss design considerations.

10

Measurement Error

The problems that error in the measurement of variables poses for identification and parameter estimation in regression analysis are reasonably well understood.[1] Our concern here is with the special problems due to measurement error *in panel models*. But to establish notation and remind the reader of results we will need in our treatment, we begin by reviewing the problem of measurement error in regression and correlation analysis.

Measurement Error in Regression Analysis

The distinction between true and observed variables is critical to the analysis of measurement error. The assumption is that observed variables are only imperfectly related to "true" variables—those that govern the causal relationships of interest. Because these true variables are only measured indirectly, they are often call "latent" variables.

To study the relationship between true and observed variables, some assumptions must be made about how these are related. Most work assumes that the true variable X_i^* and the observed variable X_i are related by the equation

$$X_i = X_i^* + e_i. \qquad (10.1)$$

---

[1] For more extensive discussion see Carter, Namboodiri, and Blalock (1975: 535–610), Duncan (1975, pp. 113–127) and Kenny (1979, pp. 74–95).

In setting the coefficient of X_i^* equal to 1, we are assuming that the scale of measurement is accurate: no systematic inflation or deflation of scores occurs. The error term e_i is assumed to have a vanishing expected value $E(e_i) = 0$ (implying that there is no *systematic* error in measurement, only random error) and to be uncorrelated with the true scores X_i^*. When more than one true score is present in a model, it will also be assumed that e_i is uncorrelated with the errors in the equations that connect the true scores. And when more than one observed variable is present, their respective errors will be assumed to be mutually uncorrelated, except where stated otherwise.

To see what effects measurement error has on regression estimates, suppose that true scores Y^* and X^* are linked by the equation

$$Y_i^* = a^* + b^*X^* + u_i. \tag{10.2}$$

We assume that Y^* is measured without error, but that observed and true scores of the independent variable are related according to Eq. (10.1). We estimate from the observed scores, $b = s_{YX}/s_X^2$. From Eq. (10.1),

$$s_{YX} = s_{Y^*X^*}$$

and

$$s_X^2 = \text{var}(X) = s_{X^*}^2 + s_e^2,$$

so

$$b = s_{Y^*X^*}/(s_{X^*}^2 + s_e^2) = (s_{Y^*X^*}/s_{X^*}^2)\{s_{X^*}^2/(s_{X^*}^2 + s_e^2)\}.$$

The factor in the brackets is defined to be the reliability of X (Jöreskog and Sörbom, 1976):

$$p_X = s_{X^*}^2/s_X^2 = s_{X^*}^2/(s_{X^*}^2 + s_e^2); \tag{10.3}$$

we consequently have $b = b^*p_X$. As p_X cannot exceed one (and usually will be less than one), measurement error will attenuate the regression coefficient by the factor p_X. Intuitively this is what we would expect: the more random error there is in observations of X^*, the more we should expect to find systematic patterns of relationships weakened.

It is straightforward to verify (by computing the appropriate variances and covariances) that the correlation between X and X^* is equal to $p_X^{1/2}$.

The *standardized* regression coefficient is given in terms of the *unstandardized* coefficient by

$$\beta = bs_X/s_Y = b^*\{s_{X^*}^2/(s_{X^*}^2 + s_e^2)\}(s_{X^*}^2 + s_e^2)^{1/2}/s_Y$$
$$= b^*(s_{X^*}/s_Y)p_X^{1/2} = \beta^*p_X^{1/2}.$$

The standardized regression coefficient is evidently also attenuated, but by a factor $p_X^{1/2}$ instead of p_X.

When Y^* is measured with error and X without error (just the opposite case of the one we have been considering), analogous reasoning shows that b is an *unbiased* estimate of b^* (no attenuation occurs). But since error in the measurement of Y inflates its variance, the standardized regression coefficient will be attenuated by a factor $p_Y^{1/2}$.

If error is present in observations of *both* dependent and independent variables, the attenuation of parameter estimates is multiplicative. For example, the standardized regression coefficient computed from observed scores will be related to the true coefficient by

$$\beta = \beta^* p_X^{1/2} p_Y^{1/2}. \tag{10.4}$$

Since the standardized regression coefficient is equal to the zero-order correlation, the latter is likewise attenuated by the product of the reliabilities in the presence of measurement error.[2]

Apart from demonstrating that measurement error can attenuate estimates of correlation and regression coefficients, these formulas can be used to derive unbiased estimates of the regression coefficients linking the true variables whenever the reliabilities of observations are known. Since the reliabilities can be computed easily provided multiple indicators of the true variables are available, this is of considerable practical value.

To see how multiple indicators permit reliabilities to be estimated, consider the model of Figure 10.1. The single true variable X^* has three indicators, X_1, X_2, and X_3. As we now want to assume that both true and observed scores are standardized, we insert the epistemic coefficients a_1, a_2, and a_3 to reconcile the scales of true and observed variables.[3] That is, we assume the factor analysis structure

$$x_j = a_j x^* + e_j, \qquad j = 1, 2, 3.$$

Since e_j and e_k are assumed to be uncorrelated for $j \neq k$, we have the following correlations among the observed variables:

$$r_{12} = a_1 a_2, \quad r_{13} = a_2 a_3, \quad \text{and} \quad r_{23} = a_2 a_3.$$

These three equations permit us to identify each a_j in terms of the observed correlations from the equation

$$a_j^2 = r_{jk} r_{jl} / r_{kl}, \qquad j,k,l = 1,2,3, \qquad j \neq k \neq l.$$

---

[2] Measurement error will also bias partial regression coefficients, but the expressions for the degree of bias are more complicated than those for simple regression coefficients; see Carter, Namboodiri and Blalock (1975, pp. 545–547); Kenney (1979, pp. 81–82).

[3] It is evident from Eq. (10.1) that without such coefficients it is not possible for both X and X^* to be standardized, for if X^* has a variance of one, X will necessarily have a larger variance.

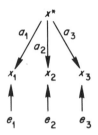

Figure 10.1 Factor structure for multiple indicators of a variable.

The sign of any one of the epistemic coefficients can be assigned arbitrarily (and ordinarily would be given a positive sign); the rest are then determined by the correlations.[4]

Since $p_{X_j} = a_j$, it follows that we can estimate the reliabilities of a single variable when three or more indicators are available.

Measurement Error in Panel Models

Along with the problem of attenuation of parameter estimates due to random measurement error, there are two special problems that rise in the analysis of panel data: the problem of unreliable change scores and the problem of autocorrelated measurement errors.

As long as we work with models that are linear in both parameters and variables, these two problems can be managed without difficulty. However, since the relationship between observed score and transformed score reliabilities has been worked out only for very simple multiplicative models (Bohrnstedt and Marwell, 1978; Taylor and Hornung, 1979; Long, 1980), it is not, in general, possible to adjust for measurement error in more complex models so readily. In the latter case, the analyst must transform the indicators of the true scores, and then compute the reliability of the transformed variables. These transformed indicators can be used in the latent variable models we analyze below.

After discussing how unreliability of change scores and autocorrelations among error terms complicate the analysis of measurement error in panel models, we turn to a discussion of specific models for measurement error in panel data.

---

[4] With four or more indicators, the measurement error model will be over-identified and LISREL can be used to obtain maximum likelihood estimates of reliabilities, as well as a global significance test of the goodness of fit of the model. It is worth noting that if X^* and Y^* are causally related, only two indicators of each variable are needed to estimate its reliability.

Unreliability in Change Scores and
Correlated Errors

Change scores are usually much less reliable than either of the two static scores that make them up. This is so because the subtraction procedure that defines the change score eliminates only the stable part of X. This means that the errors of both static scores are part of the change score.

Suppose that the observed scores X_1 and X_2 are composed of true scores X_1^* and X_2^*, and error terms e_1 and e_2. Then the observed change score is

$$\Delta X = X_2 - X_1 = (X_2^* - X_1^*) + (e_2 - e_1). \tag{10.5}$$

If the true score does not change much, the first term will be small and most of the observed score change will consist of the difference between error terms.

In the special case when e_1 and e_2 are uncorrelated, the variance of the observed change score is

$$s_\Delta^2 = s_{X_1^*}^2 + s_{X_2^*}^2 - 2s_{X_1^* X_2^*} + s_{e_1}^2 + s_{e_2}^2. \tag{10.6}$$

The reliability of the change score, p_Δ, is then

$$p_\Delta = (X_2^* - X_1^*)^2/(X_2 - X_1)^2$$
$$= (s_{X_1^*}^2 + s_{X_2^*}^2 - 2s_{X_1^* X_2^*})/(s_{X_1}^2 + s_{X_2}^2 - 2s_{X_1 X_2}). \tag{10.7}$$

A careful analysis of the final member in Eq. (10.7) shows that the reliability of the change score is determined not only by the reliabilities of the static scores, but also by the stability of the observed scores. This can easily be seen by considering the special case—one that is often a very good approximation to reality—in which the variances and reliabilities of the observed static scores of time 1 are equal to those at time 2. When use is made of these equalities, and of the identity $r_{X_1^* X_2^*} = r_{X_1 X_2}$, Eq. (10.7) reduces to

$$p_\Delta = (p_{X_t} - r_{X_1 X_2})/(1 - r_{X_1 X_2}). \tag{10.8}$$

When $r_{X_1 X_2} = 0$, the reliability of the change score reduces to the reliability of the static scores. As long as $r_{X_1 X_2}$ is positive, the reliability of the change score can be no higher than the reliability of the static score. In fact, the reliability of the change score decreases monotonically as the stability of X increases, as the reader can verify by inserting progressively higher values of $r_{X_1 X_2}$ into Eq. (10.8). Therefore, the maximum reliability of the change score is the reliability of the static scores. In most cases, the observed reliability of the change score will be less than this maximum.

A sense of how drastic the reduction in reliability can be is provided by Figure 10.2, where we have plotted values of the reliability of the change score for the special case of equal reliabilities and variances of the static scores. These calculations show that even quite reliable static scores can lead to unreliable change scores when the stability is only moderately high. For instance, when the static score reliabilities are .80 and the stability of the observed scores is .70, the change score reliability will be only .33.

As long as we have reliability estimates for the static scores it is a simple matter to estimate the reliability of the change score from Eq. (10.8). Given our earlier discussion, it would clearly be tempting to use this estimate in Eq. (10.4) to correct correlations or regression coefficients between a predictor variable and an explicitly created change score for measurement error. However, as we will demonstrate, this is an inappropriate procedure because the measurement errors for X_1 and ΔX are correlated—something not taken into account in conventional corrections for measurement error. In fact, the conventional adjustment can produce wildly erroneous estimates.

To indicate how large the error introduced by the conventional procedure can be, consider the case mentioned in the preceding paragraphs where static score reliabilities are .80 and the observed score stability is

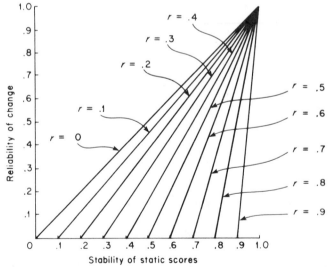

Figure 10.2.   Reliability of change scores made up of variables X_1 and X_2 as a function of static score reliabilities and stabilities. The figure is based on Table 1 of Kessler (1977b). It is assumed that the reliability of X_t is constant over time. We define $r = r_{x_1 x_2}$ to be the stability of X_t.

.70. Equations (2.2) and (2.8) can be used to calculate the correlation between the observed score X_1 and the change score; it is $-.387$. The reliability of the observed change score was already computed to be $^1/_3$. The conventional attenuation formula

$$r_{\text{true}} = r_{\text{observed}}/(p_X p_{\Delta X})^{1/2}$$

[this is Eq. (10.4) in the notation of the example] yields a true correlation of $-.75$. This estimate is in error, for the actual adjusted correlation after taking the correlation of errors into account is only $-.25$.

Before outlining the correct computation, we point out that the observed correlation between X_1 and $\Delta X (-.387)$ is *more negative* than the appropriately corrected correlation of $-.25$. This will always be the case in the absence of positive autocorrelation among the errors of the static scores.

The correct procedure is most easily explained in reference to the path diagram shown in Figure 10.3. Since the reliabilities of the static scores have been given as .80, we know that the ratio of the true score variance to the observed score variance is .80. It follows that the error term at each point in time contributes 20% of the observed score variance. The error path coefficient is consequently $(.20)^{1/2} = .447$. Similarly, the standardized path coefficient from X_t^* to X_t is $(.80)^{1/2} = .894$. We will refer to this path coefficient as the "validity" of the indicator.

Since we know that the observed correlation between X_1 and X_2 is .70,

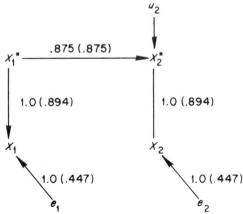

Figure 10.3. A two-wave, single-indicator measurement model, assuming independence between measurement errors across time. The numerical values are based on hypothetical data discussed in the text. Numbers in parentheses are standardized path coefficients; those outside parentheses are unstandardized path coefficients.

we can write

$$r_{X_1 X_2} = .70 = (.894)r_{X_1^* X_2^*}(.894),$$

from which $r_{X_1^* X_2^*}$ is .875. This is also the standardized regression coefficient. And since the variances of the two static scores are equal, this is also an unstandardized regression coefficient. Using Eq. (2.4a) we can obtain the regression of ΔX on X_1^* simply by subtracting 1 from this coefficient. This procedure yields a metric coefficient of $(.875 - 1) = -.125$. To obtain the *correlation* between X_1^* and ΔX^* from this regression coefficient we multiply it by the ratio of standard deviations $s_{X_1^*}/s_{\Delta X^*}$. From Eq. (2.8) we compute this ratio as 2.0 in the present example. Thus the true score correlation is $2(-.125) = -.250$.

We can see why the conventional attenuation formula fails by considering the reparametrization of our model in terms of X_1^* and ΔX^* instead of X_1^* and X_2^*, as in Figure 10.4. Here we have set the correlation between X_1 and X_1^* at .894, just as in Figure 10.3. We have also set the ratio between standard deviations of the true and observed change scores at $(p_\Delta)^{1/2} = (.33)^{1/2} = .577$.

The conventional attenuation formula estimates the true score relationship between X_1^* and ΔX^* by dividing the observed score correlation ($-.387$ in this example) by the square root of the product of the two reliabilities (in this example, we divide by the product of .894 and .577 since each of these is the square root of a reliability). As stated above, this approach yields the estimate $r_{X_1^* \Delta X^*} = -.75$.

It is clear from the diagram, though, that the conventional approach

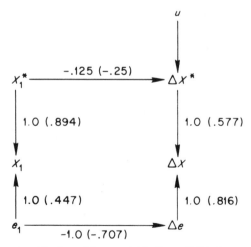

Figure 10.4. Reparametrization of the model in Figure 9.2 in terms of a change score.

fails to take account of the contribution that the correlation between the error terms e_1 and $e_\Delta = e_2' - e_1$ makes to the observed correlation $r_{\Delta X, X_1}$. To take this contribution into account, we note that the unstandardized contribution of e_1 to Δe is -1. Noting that $s_{\Delta e}^2 = e_1^2 + e_2^2 = 2s_{e_1}^2$, the standardized contribution of e_1 to Δe is $-1/(2)^{1/2} = -.707$.

Now when we use the path diagram to compute $b_{\Delta X^*, X_1^*}$ we have

$$r_{X_1 \Delta X} = -.387 = (.894)b_{\Delta X^*, X_1^*}(.517) + (.447)(-.707)(.816).$$

Solving for $b_{\Delta X^*, X_1^*}$, we find a value of $-.25$, the same estimate we obtained previously by working directly with the static scores.

Our observation in Chapter 2 that researchers working with linear models of observed scores can choose to work directly with static scores or can compute explicit change scores evidently remains true for measurement error models; here we obtained the same results either way. The choice is largely a matter of convenience. However, when working with measurement error models of the sort considered here, it is less convenient to work with explicitly computed change scores than to work with static scores because of the complications of correlated measurement errors in models with change scores. As we have just demonstrated, this is so because it is necessary to make tedious calculations in explicit change score models to adjust for the correlated errors in measurement.[5]

Measurement Error Models

We showed in the preceding section that correlations among measurement errors appear in models that contain explicit difference scores even when the errors of static score measurements are uncorrelated. In this section we consider the case where correlations among measurement errors are present even in the static score model.

As noted, the simple attenuation formula can be used to correct for measurement error only when it is assumed that the errors of measurement for the observed scores are uncorrelated. Occasions will often arise in dealing with a single variable measured at successive times when this assumption will seem unwarranted. If a particular child tends to exaggerate his or her reports in an interview on one occasion, it is not unlikely that this exaggeration will persist in later interviews. And the same can be said for a great many other sorts of errors likely to arise in the course of

[5] This conclusion is obviously predicated on our assumption that static score measurement errors are uncorrelated. One can certainly imagine models where that is extremely likely to be false, but where errors of change scores may be approximately uncorrelated with time 1 scores. In that case, the reverse conclusion would be valid.

data collection. When this is so, errors of static scores will be *autocorrelated;* that is, errors in the measurement of X at one point in time will be correlated with errors in the measurement of X at other times.

We use the term autocorrelation rather than serial correlation to accentuate the distinction between the two sorts of errors that can lead to serially correlated prediction errors. Autocorrelated measurement error refers to consistency in the determinants of erroneous reports. However, as mentioned in Chapter 7, correlated prediction errors can also be due to misspecified causal models, in which some true score common causes of X_1 and X_2 have been omitted from the model estimated. This misspecification cannot be modeled by including correlations among the measurement errors of the observed scores. It can be modeled, though, by including parameters for its effects on the specification errors in the true score model.

This chapter is concerned only with errors of measurement. The correlations among these errors can, to a degree, be estimated within the context of complex causal models. However, the number of possible correlations among errors is equal to the number of observed correlations, since each observed score presumably has some error, and this error might have a relationship to the error of some other observed scores. Estimation of all these possible correlations would completely exhaust the degrees of freedom available for model estimation. In practice, then, some restrictions must be imposed on the correlations among the measurement errors in a model.

There are three kinds of correlations among errors to consider: (*a*) autocorrelations, or correlations among the errors of identical indicators measured at different times; (*b*) "occasion" related correlations, or correlations among the errors of different indicators measured at the same time; and (*c*) "cross-test/cross-occasion" correlations, or correlations among different indicators measured at different times (Jöreskog and Sörbom, 1977).

Autocorrelated measurement errors are those that have proved to be the most troublesome in longitudinal studies. The other two classes of correlated errors are either easier to detect or appear less frequently. Occasion correlations can be detected when a postulated factor structure among observed indicators fails to reproduce the covariances among the indicators in a cross-sectional factor analysis. This sort of bias can be corrected by discarding biased indicators or by building in occasion correlations among some of the indicators. Cross-test/cross-occasion correlations usually cannot be predicted from theory, although they can be detected by studying submodels of the panel (Costner and Schoenberg, 1973) or by using somewhat more complex diagnostic tests (Sörbom,

1975). However, one should be sensitive to the fact that estimation of the minor patterns of correlation that occasionally appear here can result in overfitting the data, thereby reducing the generalizability of the model to the population.

Autocorrelated measurement errors occur frequently im empirical data and can often be anticipated on the basis of substantive considerations. Therefore they should be estimated routinely, except where extremely persuasive arguments rule them out. As long as all occasion error correlations and cross-test/cross-occasion error correlations are assumed to be absent, autocorrelated measurement errors can be estimated in all models that have two indicators and three or more waves of data, or three or more indicators and at least two waves of data. In general, though, some additional constraints must be imposed on the true score parameters or the validity coefficients to identify single-indicator m-wave measurement models (where m is an integer greater than or equal to three) or two-indicator, two-wave measurement models when they contain autocorrelated measurement errors.

In the remainder of this chapter we will study the problems of identifying and estimating measurement models. In all cases we assume that a maximum-likelihood estimation procedure of the sort utilized by the LISREL program is used to impose the consistency constraints that characterize some of the models and to estimate the parameters in over-identified models efficiently. We begin by considering models for single-indicator constructs, and then consider ways of combining measurement models with substantive models.[6] This is easy to do once the problems associated with the estimation of measurement models are spelled out, as they are in the next section of this chapter.

Single-Indicator Measurement Models

Over the last decade a number of models for estimating the reliability of a single-indicator variable measured at multiple time points have been studied. These models require, at a minimum, three waves of data for identification.

The basic path diagram for this class of models is shown in Figure 10.5. Here we have a true score X_i^* measured imperfectly at three points in time

---

[6] In practice, a measurement model will usually be part of a more complex substantive model. For example, in the path diagram in Figure 2.2, socioeconomic status and alienation are treated as true scores among which certain causal connections are posited. Each of these unmeasured true scores has some observed indicators attached to it. Assumptions about these indicators make up the measurement model. The entire model, incorporating all these assumptions, is estimated in one step (see, for example, Sörbom, 1976).

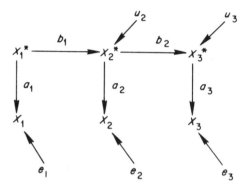

Figure 10.5. A three-wave, single-indicator measurement model.

by the indicator X_t. Measurement errors are assumed to be uncorrelated with one another. There are three correlations among the observed scores, and they can be expressed in terms of the model parameters as

$$r_{X_1X_2} = a_1b_1a_2,$$
$$r_{X_1X_3} = a_1b_1b_2a_3,$$
$$r_{X_2X_3} = a_2b_2a_3.$$

None of the individual parameters can be identified in these general expressions. However, Heise (1969) has noted that the model can be identified if one assumes that the reliabilities are constant over time, that is, $a_1 = a_2 = a_3 = a$. We then have

$$a = (r_{X_1X_2}r_{X_2X_3}/r_{X_1X_3})^{1/2},$$
$$b_1 = r_{X_1X_3}/r_{X_2X_3},$$
$$b_2 = r_{X_1X_3}/x_1X_3.$$

In three waves this model is just-identified. A fourth wave would add only one additional parameter but three correlations, thus rendering the model over-identified with 2 degrees of freedom.

A variation on this model has been proposed by Wiley and Wiley (1970), who noted that the magnitude of the error variance might well be constant over time even though the reliability (the ratio of error variance to observed score variance) changed. They proposed to estimate a metric model, then, and assume that error variance is constant. When the variances of the observed scores are constant as well, this model is identical to Heise's equal reliability model. Both models have the same degrees of freedom.

Werts and Linn (1975) proposed yet another extension of this model, this one requiring four waves of data. They noted that if the true scores are assumed to be simplex, there is no need to make any assumptions whatsoever about the reliabilities of the internal waves of the panel (waves 2 and 3 in the four-wave panel; waves 2 through $j - 1$ in the j-wave panel). However, the reliabilities and stabilities of the first and last waves are not individually identified.

To see that this is so, consider a four-wave extension of the model in Figure 10.4. We have the following expressions for the correlations among the observed variables:

$$r_{X_1X_2} = a_1b_1a_2,$$
$$r_{X_1X_3} = a_1b_1b_2a_3,$$
$$r_{X_1X_4} = a_1b_1b_2b_3a_4,$$
$$r_{X_2X_3} = a_2b_2a_3,$$
$$r_{X_2X_4} = a_2b_2b_3a_4,$$
$$r_{X_3X_4} = a_3b_3a_4.$$

If we note that a_1 and b_1 appear in these equations only together, as multiplicative factors, and similarly for b_3 and a_4, we can reparametrize the model by setting $a_1b_1 = c_1$ and $b_3a_4 = c_2$. We then have five unknown parameters, which are identified in terms of the six observed correlations as follows:

$$c_1^2 = r_{X_1X_2}r_{X_1X_3}/r_{X_2X_3},$$
$$a_2^2 = r_{X_1X_2}r_{X_2X_3}/r_{X_1X_3},$$
$$b_2^2 = r_{X_1X_3}r_{X_2X_4}/r_{X_1X_2}r_{X_3X_4} = r_{X_1X_4}r_{X_2X_3}/r_{X_1X_2}r_{X_3X_4},$$
$$a_3^2 = r_{X_2X_3}r_{X_3X_4}/r_{X_2X_4},$$
$$c_2^2 = r_{X_2X_3}r_{X_3X_4}/r_{X_2X_4}.$$

Evidently, the single degree of freedom in this model is associated with the estimate for b_2.[7] If additional waves are added to the model, other parameters become over-identifed as well.

Wiley and Wiley (1974) and Hargens et al. (1976) have developed variants on this general single-indicator approach which allow for auto-correlated errors. Given the very small margin for constraining parameter values in single-indicator models, it will come as no surprise that the assumptions required to identify parameters in these models are highly restrictive. Although these assumptions make good sense in some applications, these models should not be used uncritically in the general case. We will not review these models here.

The study of single-indicator models has been very limited. Few inves-

---

[7] When computations are done by hand, two separate estimates for b_2 are obtained, and the equality of these two estimates then tests the model. LISREL, on the other hand, provides a single best estimate of overidentified parameters, and makes possible a chi-square global goodness of fit test with degrees of freedom equal to the number of overidentifying restrictions.

tigators have used these models in empirical research, and those who have done so (Hargens *et al.*, 1976; Wheaton *et al.*, 1977) have found it necessary to introduce additional indicators to identify a plausible solution. The reason for this difficulty in single-indicator models is that they all have nearly as many, if not more parameters than correlations among the observed scores, thus making it impossible to identify a solution without imposing rather severe constraints on the model. As we will see, multiple indicator models are far more flexible.

Two-Indicator, Two-Wave Models

We begin with the very simplest of multiple-indicator models, one in which only two indicators are measured at each of two time points. The model is displayed in Figure 10.6. As there are four observed scores, there are six correlations. Assuming that measurement errors are uncorrelated, there are only five standardized parameters in the model. So the model as a whole has 1 degree of freedom.

Each parameter in the model is either identified or over-identified in the following manner:

$$a_1^2 = r_{X_{11}X_{21}}r_{X_{11}X_{12}}/r_{X_{21}X_{12}},$$
$$b^2 = r_{X_{11}X_{21}}r_{X_{12}X_{22}}/r_{X_{11}X_{12}}r_{X_{21}X_{22}}$$
$$= r_{X_{11}X_{22}}r_{X_{12}X_{21}}/r_{X_{11}X_{12}}r_{X_{21}X_{22}}.$$

The remaining coefficients, a_2, c_1, and c_2, can all be obtained from these by permuting appropriate indices. For example, to obtain the expression for c_1 start with the expression for a_1. For every set of subscripts where the first subscript is a 1, change the second subscript (from 1 to 2 or from 2 to 1, as is appropriate). The symmetry of the model tells us which subscripts to permute and which to leave alone.

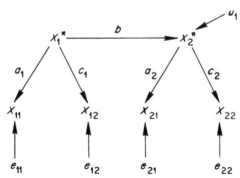

Figure 10.6. A two-indicator, two-wave measurement model. For the observed scores, the first subscript labels the wave, while the second subscript labels the indicator.

We see from these expressions that the coefficients linking the true scores to the observed scores are all just-identified, and the stability coefficient is overidentified with 1 degree of freedom.

A special case of this model is one in which the reliabilities (or in the case of metric scores, the error score variances) are constant over time; that is, $a_1 = a_2$ and $c_1 = c_2$. Each of these constraints introduces a single degree of freedom which can be used to test the assumption.

The possible existence of autocorrelated measurement errors can be evaluated globally in this model even though individual parameters for these correlations among errors cannot be estimated. Costner (1969) has shown that a special feature of the two-indicator, two-wave model is that occasion correlations among errors (e.g., between e_{11} and e_{12}) are absorbed into the validities, which means that any discrepancy between the model and observed data can be taken as an indication of cross-occasion correlated error. However, it is not possible to identify any parameters other than the true score stability coefficient in this model. Imposing consistency constraints on the validities does not help solve this identification problem.

Two-Indicator Models with Multiple Waves

The difficulty of identifying the two-indicator model when autocorrelated errors are present can be remedied by adding a third wave of data, in which case unconstrained estimates of a_t, c_t and the autocorrelations of measurement errors can all be obtained. This model is shown in Figure 10.7. Owing to the complication caused by the correlated errors, only one estimate can be obtained for b_1 and b_2. From the expressions for the correlations among the observed scores which do not contain common error terms, we obtain nine equations in the six unknowns a_1, a_2, a_3, c_1, c_2, and c_3. The six cross-time correlations can be used to solve for these six unknowns, while the three cross-sectional correlations then represent degrees of freedom for the estimates (we repeat that LISREL, being a full-information system, uses all nine pieces of information to estimate the six parameters and then provides a global goodness of fit test with 3 degrees of freedom). Four degrees of freedom are then left for use in estimating the four autocorrelated error terms in the model.[8]

In this model the epistemic parameters a_1, a_2, a_3, c_1, c_2, and c_3 and the autocorrelations of errors ξ_1, ξ_2, ψ_1, and ψ_2 can be tested for consist-

[8] The total number of degrees of freedom in the model is equal to the number of independent correlations. If there are n variables, there are $n(n-1)/2$ independent correlations. In the present case $n = 2(3) = 6$, so there are 15 degrees of freedom in all.

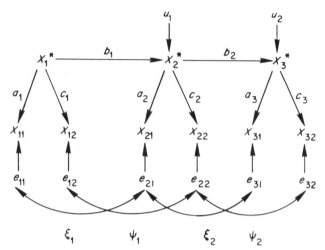

Figure 10.7. A three-wave, two-indicator measurement model with first-order autocorrelated measurement errors.

ency. That is, we can determine whether they are constant over time. We can do this by estimating a revised model in which the consistency constraints to be considered (e.g., $a_1 = a_2 = a_3$) are included and then testing the significance of the difference in χ^2 between this model and the original model with degrees of freedom equal to the number of consistency constraints imposed.

This model can be extended when a fourth wave of data is available. Here an analysis of the normal equations shows that b_1, b_2, and b_3 are each over-identified with 1 degree of freedom, that each a_t and c_t is also over-identified with 1 degree of freedom, and that the six autocorrelated errors are just-identified.

In any of these models, autocorrelated measurement errors can be estimated across three or more waves if consistency constraints are imposed on the errors for a particular indicator (e.g., $\xi_1 = \xi_2$). In two-indicator models, though, no matter how many waves of data one has available, it is not possible to estimate separate values for the correlations between error terms over shorter and longer intervals of time simultaneously.

Models Containing Three or More Indicators

We turn now to models containing three or more indicators. The simplest of these is the two-wave, three-indicator model. This model has 15 observed correlations and, assuming that measurement errors are uncorrelated, only seven standardized score parameters. The three correlations among the time 1 observed scores just-identify the three paths between X_1^* and the X_{1i}; the same is true of the three epistemic coefficients for the time 2 observations. The remaining nine cross-time correlations can be used to over-identify the parameter b, and provide over-identifying constraints on the validities as well.

It is a simple matter to incorporate autocorrelated measurement errors into this model. This can be done without making any consistency constraints on the validities.

In three-indicator models with more than two waves it is also possible to estimate unconstrained parameters for higher order autocorrelated errors (e.g., those linking nonconsecutive waves of measurement errors) without imposing any special constraints on the factor loadings (validities) or the error correlations.

These results hold, of course, for models of four or more indicators in two or more waves. In limited ways it is possible in these models to estimate correlations between errors of different indicators measured either at the same time or at different times. As we mentioned, though, it is not possible to estimate *all* of these correlated error possibilities since they equal the total number of correlations one has available for analysis. The researcher who wishes to include such correlations must decide which possible correlations to incorporate into a model, and which to exclude a priori. Some mechanical methods have been developed to aid in searching one's data for places to include such correlated errors and thereby improve the overall fit of a model to observed data (Costner and Schoenberg, 1973; Sörbom, 1975).

Measurement Error in Substantive Models

It is a simple extension of what we have been doing to incorporate measurement models as part of more complex substantive models. In the models we studied, the X_i^* were unobserved. If we include a set of such unobserved variables in one model, so that we have not only X_i^* but also Y_i^*, Z_i^* and so forth, the presumed causal relationships among these unobserved scores are interpreted as substantive relationships. In fact, after estimating a complex model of this kind it is perfectly legitimate — assuming that the model fits the data — to ignore the measurement part of the model and focus entirely on the true score parameters. This is what we did in Chapter 2 when we reviewed the model developed by Wheaton and his co-workers for the relationship between social class and alienation. These researchers estimated a complex measurement model, shown in Figure 2.2. For purposes of substantive reinterpretation of their model, though, we completely ignored the measurement component.

As long as one is willing to assume that the single-indicator scores are measured with known reliabilities, it is always possible to identify a recursive model in which only some of the variables are treated as multiple indicators of underlying true scores. For instance, Figure 10.8 shows a two-variable, two-wave model in which variable X_t^* is indicated by only one observed X_t, while Y_t^* is indicated by three Y_t. This model can be identified only if the researcher is willing to make some a priori assumption about the validities of the X_t. It is common for a researcher to do this by using several control variables that are assumed to be perfectly measured for an analysis that develops quite elaborate measurement models for the variables of primary theoretical interest (for example, see Mortimer and Lorence, 1979).

When correlated measurement errors are assumed to be absent, as they characteristically are in cross-sectional models, a measurement model for a particular true score can be identified as long as two indicators are available for that true score. In general, a measurement model can be identified for an m-variable true score model as long as $2m$ observed indicators are available, two for each of the m true scores.

The special feature of panel analysis is that in some cases it is also possible to estimate measurement models when less information is available.

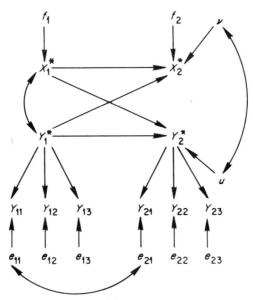

Figure 10.8. A two-wave, two-variable model with a combination of single- and multiple-indicator measurement models.

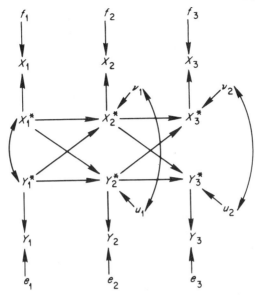

Figure 10.9. A three-wave, two-variable single-indicator panel model with lagged cross-effects between true scores.

For instance, Figure 10.9 shows a two-variable, three-wave model in which X_t^* and Y_t^* each have only single indicators. In general, we know that single indicator measurement models cannot be identified. But, as we showed in the preceding paragraphs, three-wave, single-indicator models can be identified when restrictions on the parameters of the model are imposed. This particular model is, in fact, over-identified with 2 degrees of freedom if we assume that the validities are constant.

A good deal more flexibility can be had by working with multiple-indicator models. Here, as in the measurement models for multiple indicators reviewed above, it is possible to over-identify the validities with only a few indicators in two or three waves and use these over-identifying constraints to help estimate various theoretically important patterns of autocorrelated, or more generally correlated errors. Furthermore, in all but two-wave, two-indicator models the true score parameters need not be constrained in any way to obtain identification. Therefore, it is possible to evaluate the identification of the substantive model separately from the identification of the measurement model—a feature that makes manipulation of the normal equations substantially easier than in the single-indicator case.

Overview of Results

Other than in the very simplest textbook examples of models, it will be necessary for researchers to demonstrate that a model is identified by working directly with the normal equations. Once this is known, the identifiability of the substantive model can be evaluated separately. The only time this must be done cautiously is when constraints are placed on a true score parameter in order to identify a measurement parameter, as is done in three-wave, single-indicator models (where we assumed the absence of the regression coefficient $b_{x_3^* x_1^* \cdot x_2^*}$). In cases of this sort the true score constraint cannot be considered a degree of freedom in the true score model.

Finally, when one has shown analytically that a model is identified, has estimated this model empirically, and has found that the model fits the data well, it is possible to reparametrize true score coefficients, if so desired, to study the causal structure in terms of explicit change scores and their determinants and effects. Chapter 2 outlines the procedure for doing this.

11

Alternative Approaches

The preceding chapters have outlined a strategy for estimating structural equations that represent multiwave, multivariate panel models. Several alternative methods of analyzing over-time data for individuals can be found in the literature. Here two of these, the pooling of cross sections and time series, and differential equation models, are briefly described and compared with our approach.

Pooling Methods

Suppose that T waves of observation have been collected for the variables X and Y, for N subjects, and are to be used to estimate the structural equation

$$Y_{it} = a_0 + a_1 Y_{i,t-1} + a_2 X_{i,t-1} + u_{it}. \tag{11.1}$$

The estimation could be carried out separately for each successive pair of waves, a procedure that would yield $T - 1$ separate estimates of the parameters. Or, using the methods developed in the preceding chapters, we could write separate equations for Y_2, Y_3, and so on, and constrain the coefficients of these equations to remain constant over time. This procedure will yield a single estimate of each parameter.

Pooling methods also yield a single estimate of each parameter, but the estimate is obtained through computations that differ from those we have proposed. The essence of the pooling method is to treat each wave of observations as a set of new measurements of the same set of variables. The distinction can best be seen by comparing the formats of Table 11.1 and Table 11.2. In Table 11.1 we have the conventional format for arranging scores in panel analysis. Whether we carry out separate regressions for Y_2, Y_3, and Y_4, or use the method of Chapter 3 to estimate a single four-wave panel model, we implicitly treat Y_1, Y_2, Y_3, and Y_4 as distinct variables.

Pooling methods proceed by treating each wave of observations as a set of new measurements rather than as a more extended set of data on the same observations. So, for example, a three-wave panel with 100 units of analysis will be reconceptualized as a two-wave panel of 200 units of analysis—the panel between times 1 and 2 being "pooled" with the panel between times 2 and 3 to create a doubled sample over only half as many time intervals.

We can think of this approach as involving the definition of three new variables, X_1^*, Y_1^*, and Y_2^*. The scores of X_1, X_2, and X_3 are treated as observations of X_1^*; the scores of Y_1, Y_2, and Y_3 are treated as observations of Y_1^*; and the scores of Y_2, Y_3, and Y_4 are treated as observations of Y_2^*. This reinterpretation of the scores leads to a format for the analysis of the sort shown in Table 11.2.

By combining observations in this way, the pooling method transforms a T-wave problem with observations on N cases into a two-wave panel model involving $N(T - 1)$ cases. If additional lagged effects are present in the original structural equation, these can be readily accomodated. For example, if a term in X_{t-2} is added to Eq. (11.1), the pooling method deals with successive triplets of waves, thereby transforming the problem into a three-wave panel model with $N(T - 2)$ cases.

This method implicitly assumes that the structural parameters of the equation being estimated are constant over the duration of the panel, an

Table 11.1
Format for Scores in Conventional Panel Analysis

X_1	Y_1	X_2	Scores Y_2	X_3	Y_3	X_4	Y_4
X_{11}	Y_{11}	X_{21}	Y_{21}	X_{31}	Y_{31}	X_{41}	Y_{41}
X_{12}	Y_{12}	X_{22}	Y_{22}	X_{32}	Y_{32}	X_{31}	Y_{42}
.	.	.	.	.	.	.	.
.	.	.	.	.	.	.	.
X_{1N}	Y_{1N}	X_{2N}	Y_{2N}	X_{3N}	Y_{3N}	X_{4N}	Y_{4N}

Table 11.2
Format for Scores when
Pooling Cross-Sections and
Time Series

| | Scores | |
X_1^*	Y_1^*	Y_2^*
X_{11}	Y_{11}	Y_{21}
.	.	.
.	.	.
.	.	.
X_{1N}	Y_{1N}	Y_{2N}
X_{21}	Y_{21}	Y_{31}
.	.	.
.	.	.
.	.	.
X_{2N}	Y_{2N}	Y_{3N}
X_{31}	Y_{31}	Y_{41}
.	.	.
.	.	.
.	.	.
X_{3N}	Y_{3N}	Y_{4N}

assumption that can be evaluated by first estimating a multiwave consistency model. If this assumption is incorrect, pooling is not appropriate.[1] If the assumption is correct, and if serial correlation among the disturbances is absent, conventional OLS methods can be used to estimate Eq. (11.1) without bias. Most work on pooling methods, though, has involved the more difficult case where the assumption that error terms are serially uncorrelated cannot be made. Indeed, as will be discussed, most complica-

---

[1] Practitioners have not always heeded this admonition. Stevenson (1972), for example, pools cross-sections and time series to determine the impact of external investment on economic growth in Latin America. After first computing cross-sectional correlations of various economic indicators for the years 1961–1967, he pooled them to estimate overall correlations for the period. Since the cross-sectional correlations vary tremendously over the seven years (e.g., the cross-sectional correlation between direct investment and economic growth ranged between −.72 and +.62), and Stevenson suggests reasons for thinking that the causal relationships were not invariant over the period, the pooled estimates are probably meaningless. By contrast, Papanek (1973), who investigated similar questions with a somewhat larger sample ($N = 34$ for the 1950s, $N = 51$ for the 1960s) first estimated his regressions using a dummy variable for decade. When the coefficient for the dummy proved to be statistically insignificant, he estimated separate regressions for the two decades. Since the coefficients for the two equations were similar, he proceeded to pool them.

tions in the estimation of pooled models arise from serially correlated errors.

The decision whether or not to pool one's data is largely dependent on the relative importance of increasing the size of the sample and increasing the number of time points in the panel. Pooling methods are clearly preferable to other approaches when N is very small, as it will be in the comparative analysis of nation-states or multinational corporations. Here the pooling of observations across multiple points in time—whether panel models or cross-sectional models are subsequently estimated—is required to obtain parameter estimates with a meaningful degree of precision. Stevenson (1972), for example, pooled 7 years of data on 7 nations to determine the impact of external investment on economic growth in Latin America. By pooling he effectively increased his sample size to 49, with a substantial reduction in the standard error of his parameter estimates.

When N is sufficiently large that large-sample properties of the covariances can be inferred, the use of pooling is less desirable. This is so because pooled models treat long panels as nothing more than a succession of short panels. Thus information on long lag covariances is thrown away. As we have shown in various chapters above, this long lag information can be extremely useful for estimating otherwise intractable models via the imposition of consistency constraints (Chapter 3). It can be used to estimate the magnitude of serially correlated errors (Chapter 7). It can also be used to identify single-indicator measurement models (Chapter 10).

When a researcher is fortunate enough to be working with a panel of many waves, the loss of long lag information through pooling will be of no real importance. Adjacent sets of waves can be pooled in an instance of this sort, yielding a multiwave panel with a substantially increased sample size and, with this, a much reduced standard error of the parameter estimates.

However, instances of this sort are rare and are usually confined to archival research. In the more characteristic case, T is small (between 2 and 4 time points) and N is at least moderately large. Here pooling is much less attractive than a multiple-wave approach, since the flexibility gained by having T equal to, say, 3 or 4 instead of 2 is much more valuable than the precision gained by increasing the sample size to $2N$ or $3N$.

Serial Correlation in Pooled Models

As noted, the estimation problems that arise in multi-wave pooled models are no different from those that occur in other multi-wave models. When the researcher is fortunate enough to have 10 or 15 waves of data, it is a simple matter to pool adjacent series of 4 or 5 waves and still inflate the sample size two- or threefold. Error correlations are somewhat more complex than in panel models which have not been pooled, since the possibility of correlated error across units of observation arises from the fact that each unit of analysis is treated as $T - 1$ observations. But, as will be discussed, this complication can be easily handled by introducing dummy variables for each unit of analysis or by expressing observed scores as deviations from their intraunit means (i.e., the means of the different observations for a single case).

A much more difficult situation arises when the data have both a small N and a small T. Here there is no choice but to pool the data into a series of $N(T - 1)$ two-wave observations. Unfortunately, the management of serially correlated errors is especially difficult when working with only two waves.

The bulk of work on pooling methods has involved the particularly difficult case where only two waves of data are available for analysis and errors are serially correlated. Most forms of serial correlation must be estimated here with the instrumental variables technique discussed at the end of Chapter 7, a technique that is much less efficient than those available in the multiwave case. In a single-equation system, like that in Eq. (11.1), this technique entails using X_{t-1} as an instrument for Y_{t-1} in the equation for Y_t. In a simultaneous equation system, where X and Y are mutually dependent, X_{t-1} cannot be used as an instrument since it is no longer uncorrelated with u_t. Thus, some third variable Z must be used instead. In practice, it can be difficult to find a Z of this sort.

In the special case when serially correlated errors are assumed to be constant at all orders,[2] it is possible to use an estimation procedure that regains the efficiency available in multiwave estimation procedures. This procedure hinges on the fact that each individual unit of analysis in a pooled model is treated as $T - 1$ observations. If serial correlation is constant at all orders, these multiple observations of each unit i will contain a constant error component, u_i. We can then write the error for each unit at time t as

$$u_{it} = u_i + v_{it},$$

where v_{it} is a random error component uncorrelated with Y_{t-1}. Then we

[2] We have previously discussed the implications of these assumptions about error terms in Chapter 7. The only circumstance where the error terms are likely to have the assumed structure is when all variables that have been omitted from the model are constant with time. Should one have reason to suspect that this is untrue, the corresponding structure of the error terms should not be assumed. With a different assumed error structure, of course, the method described in the text cannot be carried out.

can rewrite Eq. (11.1) in the following equivalent form:

$$Y_{it} = a_0 + a_1 Y_{i,t-1} + a_2 X_{i,t-1} + D_i u_i + v_{it}, \qquad (11.2)$$

where D_i is a dummy variable for the ith individual unit of analysis.[3]

By including a set of $N - 1$ dummy variables of this sort (one for each unit of analysis, with the indeterminancy eliminated by omitting one of the N dummys) into a pooled regression analysis, the structural parameters a_1 and a_2 can be estimated without bias. The same result can be achieved by eliminating the intraunit constants prior to estimation, expressing the observed scores as deviations from their intraunit means.

The efficiency of this estimate can be improved by recovering the value of the serial correlation from the estimated D parameters and using this to obtain a generalized least squares solution. A particularly appealing feature of GLS when used in this way is that the relative weights of N and T in the estimation are fixed in a conceptually clear way (Hannan and Young, 1977). Although there are $N(T - 1)$ observations in the pooled model, this information does not intuitively seem as valuable as it would were it obtained from $N(T - 1)$ independent two-wave observations. But just how valuable is it? Generalized least squares answers this question by weighting in terms of the magnitude of the serial correlation coefficient. When this coefficient is perfect, successive observations of the same unit are completely dependent and the sample size is effectively N. The generalized least squares estimator in this case averages successive observations *within* units and relies on variation among these averages *across* units to obtain the parameter estimates. In the more realistic case where the correlations among errors lie between 0 and 1, the procedure weights cross-sectional and temporal variation in the variables inversely proportional to the correlation. Thus the more independent the observations in successive waves are, the more heavily they count in the analysis. In the extreme, when the correlation vanishes, all observations are treated as independent and the relative weights given to cross-sectional and time-series variation are, respectively, N and T.

Unfortunately, in the presence of mutual dependence between X and Y, this dummy variable GLS approach fails and an instrumental approach must be employed. Hannan and Young (1977) also demonstrate the usefulness of combining the use of an instrumental variable with the dummy

---

[3] Some of the literature also treats the case where a time-specific error term u_t^* is present. Such a term represents effects that are common to all units of analysis, but that vary with time. By grouping such a term with the intercept a_0 in Eq. (11.1), we see that in effect it represents a time-dependent intercept. As Hannan and Young (1977) note, no fundamental analytical difficulties are posed by such terms as long as they have the same properties as the v_{it}; consequently, they will be ignored in our discussion.

variable approach in a situation of this sort. Neither of these approaches, though, is as efficient as the elimination and differencing approaches that are available when three or more waves of data are available. Thus we see that the estimation of serially correlated error is more problematic in two-wave pooled models than in multiwave models. In practice, this is an extremely damaging weakness, one that makes the use of pooled models something that we would advocate only in the extreme case where N is so small that no choice exists but to pool observations in an effort to obtain stable parameter estimates.[4]

Differential Equations

Differential equations are an especially convenient mathematical tool for studying change. They are not widely used in the social sciences because their application is technically restricted to variables that change smoothly (continuously). Most of the variables of interest in the social sciences can only take discrete values and, consequently, are not continuous. Nevertheless, differential equations may provide very good approximations for the problems we encounter. Where this is so, they can be a powerful analytical tool.

Here we will review the use of differential equations for the study of change. Space limitations preclude our presenting a full exposition of the theory of differential equations.[5] Instead, we aim at the more modest goals of enabling readers to understand the main ideas behind the application of differential equations to the analysis of change, and how this approach differs from those we have been considering. The reader who understands this material will be in a position to follow the work of those who use differential equations, to communicate with experts regarding applications to their own work, and to undertake further study.

Before defining a differential equation, we must first define a derivative. The derivative of a variable Y with respect to time t is the instantaneous rate of change of Y. Roughly speaking, it measures the amount by which Y changes when t changes by an infinitesimal amount. Some examples will make clear what this means. If Y is the distance a car has traveled along a

---

[4] For readers who believe, on the basis of this consideration, that pooling is required for a particular problem they are addressing, a fuller discussion of these methods can be found in Hannan and Tuma (forthcoming).

[5] More extended treatment of differential equations can be found in Coleman (1968), Nielson and Rosenfeld (1978), Greenberg (1979), and in the many standard mathematics texts on differential equations, Kaufman (1976) has pointed out that one of Coleman's equations is in error.

road, then its derivative with respect to time at any moment is the rate at which the car's position is changing. We all recognize this to be the speed of the car at the given moment (e.g., 37 miles per hour). If Y is population, then its time derivative is the rate at which the population is changing (e.g., by so many thousands of people per year). Although we can define derivatives with respect to other variables than time, we shall have no need for them here.

The derivative of Y with respect to t is conventionally denoted by the symbol dY/dt. The important point to remember here is that d is *not* an algebraic quantity. It cannot be cancelled in this expression. The symbol dY/dt should be treated as a whole, and read as an imperative: take the derivative of Y with respect to t! We note that the derivative is a signed quantity; it can be positive or negative. If Y has a negative derivative, it simply means that Y is declining with time.

A differential equation is an equation that contains the derivatives of one or more dependent variables with respect to one or more independent variables. We would ordinarily write a differential equation if we want to express the dependence of the change in some quantity on other variables. Does the rate of population growth in a society depend on its level of economic development? If so, we might write an equation expressing the time derivative of its population in terms of its gross national product. Is income growth influenced by present levels of personal income? If it is, we would write an equation expressing the derivative of income with respect to time as a function of current income.

Since the derivative of a variable with respect to time measures its rate of change, dynamical systems are quite naturally represented by differential equations. It is thus hardly surprising that researchers should turn to differential equations when they want to analyze panel data.

Differential equations can take many forms, depending on the dynamical properties of the system being modeled. For example, if the rate of change in Y is constant with time, we would write

$$\frac{dY}{dt} = c = \text{constant.} \qquad (11.3)$$

If the rate of change in Y is assumed to depend linearly on Y itself and on an exogenous variable Z, the appropriate differential equation would be

$$\frac{dY}{dt} = a_0 + a_1 Y + a_2 Z. \qquad (11.4)$$

Just as systems of simultaneous linear regression equations can be written to express the reciprocal dependence of two variables in the regression formalism, here we can likewise write systems of linear equations to express reciprocal influence. For example, the set of linear equations

$$\frac{dY}{dt} = a_0 + a_1 Y + a_2 X, \tag{11.5}$$

$$\frac{dX}{dt} = b_0 + b_1 X + b_2 Y \tag{11.6}$$

corresponds to the assumption that the rate of change in Y depends linearly on the values of X and Y, and similarly for the rate of change in X. When both equations are valid simultaneously, X and Y will depend on each other; whereas Z can be treated as exogenous in Eq. (11.4), X cannot be treated as exogenous in Eq. (11.5).

Standard methods exist for solving such equations.[6] Solutions take the form of an explicit functional dependence of the dependent variable on the independent variables, the parameters of the differential equation, and the initial values of the variables in the equation. Thus, if the clock starts running at $t = 0$, and $Y = Y(t)$ is the value of Y at time t, the solution of Eq. (11.3) is $Y = Y(0) + ct$. The linear dependence of the solution on time is just what we would expect from a differential equation that asserts that the rate at which Y is growing is a constant.

For Eq. (11.4) the general solution can be shown to be

$$Y = Ce^{a_1 t} - (a_0/a_1) + a_2 e^{a_1 t} \int Z e^{-a_1 t} \, dt. \tag{11.7}$$

Here C is an arbitrary constant and $\int$ is the integral operator (the reader who has not encountered it before should not panic; it is simply the inverse operation to taking the derivative; it has the same relation to taking the derivative as adding does to subtracting). Roughly speaking, it denotes an infinite sum of infinitesimal quantities.

The variable Z in Eq. (11.4) has already been specified to be exogenous; its time dependence is not determined by the differential equation, but must be given. For any specified time dependence of Z, the integration indicated in Eq. (11.7) can be carried out to obtain an explicit time dependence for Y. For example, if Z is proportional to e^{-bt} (a declining exponential if b is positive), the final term in Eq. (11.7) will itself be proportional to e^{-bt}, if Z oscillates sinusoidally, so will the final term in Eq. (11.7). Readers who have already mastered integral calculus will be able to verify these statements; those who have not will have to take them on faith.

Note that if Z vanishes entirely, Y approaches the quantity $-a_0/a_1$ asymptotically at large time. Referring to Eq. (11.4), we see that this is the value of Y for which $dY/dt = 0$. Thus it is a steady-state value for Y.

---

[6] See the sources listed in footnote 8.

An Example from Classical Mechanics

Sociologists accustomed to working with structural equations may find solutions of this form unfamiliar. Structural equations model the functional dependence of the variables in a system on one another. For example, they might specify the dependence of X and/or Y on Z, but ordinarily would not specify an explicit time dependence of X and Y. By contrast, these differential equations specify how the *changes* in some variables depend on other variables, and the solutions yield *explicit* time dependences for these variables.

This distinction can be clarified by considering a simple example from classical mechanics. Suppose we throw an object horizontally along a frictionless floor. If we use x and y to denote the position of the object along mutually perpendicular axes, the differential equations governing the motion of the object after it has been released will be

$$dx/dt = v_{0x}, \; dy/dt = v_{0y} \tag{11.8}$$

where v_{0x} and v_{0y} are, respectively, the x and y components of the velocity at the time the object is released. Together, these two equations assert that the x and y components of the object's velocity are each constant, and equal to the same components at the time the object is released. Since there is no friction, there is nothing to slow the object down or deflect it.

If the object is released at time $t = 0$ with initial position $(x,y) = (x_0, y_0)$, the solutions to these equations are

$$x = x_0 + v_{0x}t, \quad y = y_0 + v_{0y}t. \tag{11.9}$$

These equations specify the position of the object at any time t. If we now solve the first of Eqs. (11.9) for t and substitute the expression we obtain in the second of Eqs. (11.9), we obtain a linear equation

$$y = (y_0 - x_0 v_{0y}/v_{0x}) + (v_{0y}/v_{0x})x. \tag{11.10}$$

Note that time has been eliminated from Eq. (11.10). In place of two equations specifying the time dependence of x and y, we have a single *static* equation for the trajectory of the object, specifying the functional dependence of y on x. Since this dependence is linear, the object moves in a straight line across the floor.

Whenever we have parametric equations of the form (11.9) (so-called because x and y are given in terms of the parameter t), we can eliminate time to obtain an equation for the trajectory $y = y(x)$. We cannot, however, go in the opposite direction. A given trajectory is compatible with many different types of time dependence for x and y.

The regression equations we commonly encounter in sociology are analogous to static trajectories. They tell how one variable depends on other variables, but lack dynamical information. Implicitly, they assume that the system under study is in equilibrium. To see this, set $dY/dt = 0$ in Eq. (11.4), and solve for Y. The resulting expression

$$Y = -a_0/a_1 - (a_2/a_1)Z$$

is the form of functional dependence we typically find in a regression equation.[7] If an explicit time dependence of Z is given, this equation will yield a time dependence for Y but the dependence so obtained will be valid only after equilibrium has been reached.[8]

Differential Equations and Panel Models

In empirical research, one proceeds by setting up differential equations that express the causal relationships assumed to govern the evolution of the system under study. After solving the differential equations, the parameters can be estimated by fitting the solutions to observed values of X and Y. This approach can be used to analyze panel data, since observations on one or more variables are collected at successive points in time.

---

[7] For simplicity the stochastic component, or disturbance term that appears in a regression equation, will be omitted from the differential equations we consider. Since it is the expected value of the solutions that will be fit by the data, this omission poses no problems as long as error terms are well-behaved (that is, they are homoscedastistic and mutually uncorrelated). Care is needed when the differential equations that model a system are not solved directly, but are manipulated before being solved, since these manipulations may lead to error terms that violate these assumptions. For example, the coupled first-order linear differential equations (11.5) and (11.6) can be solved by analogy to the method by which the coupled difference equations in the Appendix to Chapter 8 were solved. This procedure transforms the original equations into a pair of uncoupled second-order linear differential equations. It is too often forgotten that even if the error terms in the original equations are well-behaved, those in the uncoupled equations will be serially correlated. Blumstein, Cohen, and Nagin (1976) neglect this in their analysis of time series for imprisonment rates in different societies.

[8] Sorenson (1973) has explored some of the ramifications of this relationship between differential equations and regression equations. He points out that if a system that has not yet reached equilibrium is studied with the regression equation (implicitly assuming that equilibrium has been reached), the relative magnitude of the constant term in the regression to the other terms will depend on time. Even when equilibrium has been reached, the cross-sectional estimation of the regression equation can be misleading when used to compare estimates across populations. A given regression coefficient will depend on both a_2 (say) and a_1. Thus coefficients in different populations can differ even when the corresponding dynamical parameters a_1 are the same for the two populations. This problem could distort a comparison of the influence of background variables on income or status attainment between different groups (such as male and female, or black and white). Within a single population, this problem does not arise: in equilibrium, the relative magnitudes of the regression coefficients will be the same as those in the differential equation. However, no separate estimate of the parameter a_1 is possible.

Hannan and Tuma (1979) have recently argued that by comparison with other methods for analyzing panel data,

> A major advantage [of using differential equations] is that it makes the timing between waves irrelevant . . . Thus, for at least the class of linear differential equation models, the identification problem [having to do with the correct choice of causal lag] does not arise.

Referring to our Eqs. (11.5) and (11.6), Hannan and Tuma assert that "Inspection of the [functions that solve these equations] shows that the spacing of observations is taken into account in a perfectly natural way."

As we will see, differential equations do offer certain advantages over other methods for analyzing over-time data, but this is not one of them. Referring to Eqs. (11.3)–(11.6), the first thing we note about them is that in each case the rate of change in a variable depends only on the *instantaneous* values of the right-hand members. In other words, the equations have *implicitly specified that lagged effects are absent*. Since one can also assume that lagged effects are absent using regression equations, differential equations offer no particular advantages in this respect.

It is possible to introduce lagged effects explicitly into a differential equation. It might make sense, for example, to assume that change in Y depends on the instantaneous value of Y, but that the effect of Z on Y is lagged. If the appropriate lag is τ, then instead of Eq. (11.4), we would have the differential equation

$$\frac{dY}{dt} = a_0 + a_1 Y + a_2 Z(t - \tau). \tag{11.11}$$

If τ is known, its value can be inserted in the equation; if not, it becomes an additional parameter to be estimated from the data. This estimation can be carried out for any given dependence of Z on time, except that τ cannot be identified uniquely if Z declines exponentially with time, or is a periodic function.[9]

---

[9] To see this, note that in the former case $Z = Z_0 e^{-kt}$. If the effect of Z is lagged by τ, the contribution Z makes to the differential equation will be $a_2 e^{k\tau} Z_0 e^{-kt}$. The effect of τ is here absorbed into the effect of a_2. In the latter case, suppose Z to be periodic, with period T. This means that $Z(t + T) = Z(t)$. Consequently, the effect of $Z(t - \tau)$ cannot be distinguished from the effect of $Z(t - \tau - T)$.

A more realistic model of lagged effects might assume a continuously distributed lag. This can be accomodated by replacing the term $a_2 Z(t - \tau)$ in Eq. (11.11) with the more complicated expression

$$a_2 \int_0^t Z(t - \tau)\phi(\tau)d\tau.$$

Here ϕ is a continuous weighting function. Differential equations involving terms of this kind

In the same way, lags can be introduced into the set of coupled Eqs. (11.5) and (11.6) as follows:

$$\frac{dY}{dt} = a_0 + a_1 Y + a_2 X(t - \tau),$$

$$\frac{dX}{dt} = b_0 + b_1 X + b_2 Y(t - \tau).$$

These equations, too, can be solved, and the solution will depend on τ as well as on the six parameters a_0, a_1, a_2, b_0, b_1, and b_2. One need know nothing about differential equations to realize that observations on two variables at two points in time will not be capable of providing unique estimates of seven parameters (six if the lag is specified by the researcher) just because the functions being fit are solutions to differential equations. In terms of estimating parameters, differential equations are no different in their requirements than structural equations for panel data.

In addition, the researcher who collects observations at grossly inappropriate lags runs exactly the same danger of misinterpreting the causal processes at work when the data are analyzed using differential equations as when using regression equations. If the waves are too far apart, the dynamical aspects of the process will be overlooked, and the researcher will see only a succession of equilibrium states. The panel design will offer nothing that a cross-sectional design will not provide more cheaply. If the waves are collected at lags spaced too closely, the process of interest will not have had time to unfold. With observations collected daily, little would be learned about the decline and fall of the Roman Empire!

The relationship between differential equations and structural equations can be seen more clearly if we return to the structural equations we considered in chapter 2. There, we pointed out the equivalence between the equation

$$Y_2 = b_1 Y_1 + b_2 Z \tag{11.12}$$

and the equation

$$\Delta Y = Y_2 - Y_1 = (b_1 - 1)Y_1 + b_2 Z. \tag{11.13}$$

---

can be solved using Laplace transforms (see Greenberg, 1979). We regard models of this kind as more realistic than those that assume discrete lags, because most social processes involve influences that are at work continuously over some period of time, rather than coming into play only at discrete moments. Where this is the case, a panel model based on discrete lags approximates the influences actually at work: a single lagged effect will summarize the net effect of influences that occur both before and after the moment at which observations are made. However, as noted earlier, bias due to temporal misspecifications can arise through this approximation.

With the scale of time such that the interval Δt between waves is treated as one time unit, this is

$$\frac{\Delta Y}{\Delta t} = (b_1 - 1)Y_1 + b_2Z. \tag{11.14}$$

The quantity on the left, $\Delta Y/\Delta t$, is the ratio of the change in Y over some period of time, to the duration of time over which that change has taken place. The ratio is, therefore, the average rate at which Y has changed during the time period in question. As the time period Δt approaches zero, the left-hand member of Eq. (11.4) becomes the instantaneous rate of change in Y, which we know is the derivative of Y with respect to t. Equation (11.14) in this limit then becomes the differential equation (11.4) (without the constant term in the right-hand member). This shows that a regression equation can be translated directly into a differential equation as long as the interval between waves is small (on a time scale set by the processes under study).

To go in the opposite direction, from differential equation to difference equation, a complication arises. If we wish to translate the differential equation

$$\frac{dY}{dt} = aY$$

(with disturbance term omitted) into a difference equation, we clearly should replace dY/dt with $Y_2 - Y_1$. But should Y be replaced by Y_1, Y_2, or perhaps be the average of these? In the first instance we derive the regression equation

$$Y_2 = (1 + a)Y_1,$$

in the second, the equation

$$Y_2 = Y_1/(1 - a),$$

and in the third, the equation

$$Y_2 = [(1 + a/2)/(1 - a/2)]Y_1.$$

The optimum choice is a technical question that is still under study (Bergstrom, 1976); here we only note that the differences among the coefficients are of order a^2. As long as a is small, these differences will be small. The parameter a, we note, is an inverse measure of the time scale of the growth process described by the differential equation. This finding, then, corroborates our earlier assertion that as long as the interval between waves is small compared to the time scale of the process under investigation, differential equations and regression equations are equivalent.

To avoid these ambiguities, we recommend that whenever possible, the translation of differential equation to regression equation be done by *solving* the differential equation. To illustrate this procedure, we continue to work with Eq. (11.3), with the parameter a_0 set equal to zero for simplicity of exposition. If Z is constant, and the first and second waves are collected at times t_1 and t_2, the solution to the differential equation is

$$Y_2 = e^{a_1(t_2-t_1)}Y_1 + (a_2/a_1)[e^{a_1(t_2-t_1)} - 1]Z. \qquad (11.15)$$

If the third wave is collected at time t_3, the equation giving Y_3 in terms of Y_2 is, by analogy,

$$Y_3 = e^{a_1(t_3-t_2)}Y_2 + (a_2/a_1)[e^{a_1(t_3-t_2)} - 1]Z. \qquad (11.16)$$

Provided observations have been collected at equal intervals, we can set $t_3 - t_2 = t_2 - t_1 = T$. Setting

$$b_1 = e^{a_1 T} \quad \text{and} \quad b_2 = (a_2/a_1)(e^{a_1 T} - 1), \qquad (11.17)$$

we regain Eq. (11.12). This demonstrates that when the observations collected at equally spaced intervals are generated by an underlying differential equation of the form (11.12), the parameters of the equation can be estimated through structural equations of exactly the sort we have been considering.

When observations are collected at intervals that are *not* equally spaced, the solutions to the differential equations tell us how the parameters of the corresponding structural equations depend on time. Hannan and Tuma (1979) correctly note that this information facilitates the comparison of parameter estimates where systems are assumed to be governed by the same sort of dynamics, but observations are collected at different intervals. Thus, if the interval between the first and second observations is one time unit, while the interval between the second and third observations is four time units, the coefficients expressing Y_2 in terms of Y_1 and Z are

$$b_1 = e^{a_1} \quad \text{and} \quad b_2 = (a_2/a_1)(e^{a_1} - 1),$$

while those expressing Y_3 in terms of Y_2 and Z are

$$b_1^* = e^{4a_1} \quad \text{and} \quad b_2^* = (a_2/a_1)(e^{4a_1} - 1).$$

As long as a_1 is small, we can keep only the lowest order terms in the series expansion $e^x = 1 + x + x^2/2 + \ldots$. Expanding the expressions for b_1 in this way, we see that $b_1 = 1 + a_1$, while $b_1^* = 1 + 4a_1$; thus the difference is of order a_1. On the other hand, a similar coparison of b_2 and b_2^* yields the prediction that b_2^* is four times as large as b_2. We would thus expect the cross-coefficient expressing the influence of Z on Y in a structural equation to change more drastically than the stability coefficient when lags between successive observations are of different lengths.

To illustrate this procedure, we return to the work of Wheaton *et al.*, discussed previously in Chapters 2 and 8. If we re-estimate the path model of Figure 2.3 on the assumption that A_1 is uncorrelated with the disturbance of A_3, we obtain the following regression equations:

$$A_2 = .712A_1 - .183SES + \text{error term,} \qquad (11.18)$$

$$A_3 = .514A_2 - .335SES + \text{error term.} \qquad (11.19)$$

Comparing Eq. (11.18) with (11.17), we estimate

$$a_1 = \ln b_1 = \ln (0.712) = -.34,$$

and then estimate

$$a_2 = a_1 b_2/(e^{a_1} - 1) = -.34(-.183)/(.712 - 1) = -.216.$$

A similar computation for Eq. (11.19) yields the estimates $a_1 = -.166$ and $a_2 = -.114$.

Comparing the two sets of parameters, we see that a_1 is smaller in magnitude over the time span linking the second wave with the third than it is over the time span linking the first wave with the second, but in both instances it is negative. This means that the "regression to the mean" effect declines somewhat over time. The tendency of people with exceptionally high levels of alienation to be declining in their alienation is not as high at later times as it is at earlier times. The cross-coefficients at both times are negative, but in the later period the effect is only about half as large as in the earlier period.

The conclusions reached here are in full agreement with those obtained in Chapter 8, where we assumed that the 4-year data (the third wave) were generated from the second wave by a structural equation in which the true lag was 1 year.

Differential Equations and Continuous-Time Models

Where differential equations hold the clear advantage is when the data are not panel data at all. In those cases where variables can change their levels at any time, and this time is known, the solutions to the differential equations can be estimated directly, without the aggregation over time that would be required to estimate a panel model. Since aggregation of this kind can lead to temporal misspecification (as pointed out in Chapter 5), it is to be avoided where unnecessary.

For most of the variables of interest to social scientists, records of the levels of these variables are not kept continuously. In studying public opinion, for example, it is not too hard to ask questions from time to time, but a *continuous* measurement would be virtually impossible. In other sorts of analyses, we are given only data that have already been aggregated (e.g., annual birth rate).

Where the investigator can gain access to the appropriate archives, this need not be a limitation. If the time of every birth, death, and immigration to and from a city were recorded, a continuous measurement of the city's population would be available. In some populations, a continuous record of subjects' incomes could be constructed. The records of a psychiatric institution will record dates of admissions and releases of inmates that could presumably be culled by a researcher. So if one wished to study the effect that the size of an institution's population has on its suicide rate, this could be done using differential equations instead of by estimating a structural equation in which the average annual population (say) is used to predict the annual suicide rate.

This approach converges on approaches based on continuous-time stochastic models.[10] Stochastic models are probabilistic models for transitions between states of a system. Given an initial state, a stochastic model will predict the probabilities of the system being found in any given state at a later time. When transitions can occur at any time whatsoever, rather than only at discrete instants, we have a continuous-time stochastic process.

If the state of a system is specified by the number of events of a particular kind that have occurred (e.g., the number of riots in a city), the mean rate of events becomes an important parameter of the model. This rate is an interval-level variable, and can thus be regressed on other variables, or used to predict other variables.[11] Thus a panel analysis of rates can be joined to a stochastic model for events.

These two approaches compliment one another. Analysis of stochastic models can help to locate such effects as reinforcement and contagion, while the causal analysis of model parameters is the surest way of identifying heterogeneity in the stochastic model, as well as being of interest for its own sake.

---

[10] Treatments of stochastic models accessible to social scientists can be found in Coleman (1964), Fararo (1968, 1973), Bartholomew (1973), Singer and Spilerman (1976), and Greenberg (1979).

[11] This approach is applied to urban race riots by Spilerman (1971).

12

Design Considerations

The preceding chapters have largely been concerned with methods for analyzing and interpreting panel data that have already been collected. Here we turn our attention to considerations that arise in the data collection phase of research. We will discuss such questions as: Should one perform a panel study, or will some simpler research design suffice? How much time should separate the waves of a panel study? How can one minimize sample attrition, and adjust for attrition that has occurred?

Only to a limited degree are these statistical questions. To a great extent they concern pragmatic questions, such as how to locate the subjects of a panel study in follow-ups. Our emphasis will be on strategies researchers have developed for dealing with such questions in their own work.

Designing a Panel Study

Compared to other sorts of data utilized by the social scientist, panel data can be expensive to obtain. In addition, analysis of the data cannot begin until at least two waves of data are in hand. In some studies this could be a matter of years. These drawbacks should make the researcher think about two important issues before planning a panel study. First, the possibility of using cross-sectional or trend data should be considered carefully. If a causal model can be plausibly identified with cross-sectional data, then a panel study may be unnecessary. Additionally, if

one is not concerned with questions of causality at all, but only with measuring aggregate change in individual variables in a population, then trend
data will prove adequate.

If it is decided that only panel data will suffice for the problem at hand,
a second issue should be addressed: whether a prospective panel study
should be launched, or other alternatives adopted. The foremost alternative to be considered is clearly the secondary analysis of an existing set of
panel data—either data explicitly collected by someone else, or data that
can be culled from archives. But two others—follow-back panels and
catch-up prospective panels—deserve consideration as well. Since they
are somewhat less well-known to sociologists, they will be discussed here
in some detail.[1]

The Follow-Back Panel

Follow-back, or restrospective panels select a cross-sectional sample in
the present and then utilize archival data from an earlier time point for
each unit of analysis to create the through-time feature of the study. A
catch-up panel involves the selection of a cross-sectional sample from an
archival source at some time in the past, and then locates the units of analysis in the present for subsequent observation. Both designs have advantages over prospective panel studies if the archives contain information
on the variables of interest to the investigator.

Let us first consider the advantages and disadvantages of the retrospective panel. Through a retrospective panel, changes that have taken place
over a great many years can be studied in a short amount of time.
Follow-back studies can also be inexpensive, especially when conducted
entirely on the basis of archival data. Robins (1979b), for example, de-
scribes a study aimed at understanding the childhood predictors of mental
disorder in which psychiatric patients were located on the basis of hospital records and followed back to school records, which were used to
match them with classmates who (presumably) were not mental patients
as adults.

Of course, follow-back studies can also be based on a current survey.
In this format, respondents are asked about past school attendance, and
then early school records are obtained on the basis of these leads.

The follow-back design is particularly advantageous in studying a rare
population, because it permits the researcher to assemble a sample

[1] This discussion draws heavily on Robins (1979a,b).

without having to begin with an enormous baseline sample at an earlier time. This feature makes the follow-back study a popular design in psychiatric epidemiology. A typical study might start with a group of identified schizophrenics who can be compared on some series of childhood characteristics with a sample of normal respondents interviewed at the same time. By contrast, a prospective panel study would require a very large sample of children to guarantee having enough adult schizophrenics in the sample to make a comparison possible.[2]

Despite its advantages, the disadvantages of the follow-back design are such that a researcher who is not studying a rare population may not want to use it. For one thing, all follow-back studies must rely on earlier archival data. In the United States, data of this sort are usually limited to school, hospital, and military service records. With the exception of military records, these sources are likely to be of variable quality depending on the respondent's social class background and the part of the country in which he or she was raised.[3]

Another disadvantage of the follow-back design is that it can yield a censored sample, one in which a unit of analysis is excluded from the sample because an event that has occurred has not been observed. Duncan (1966b), for example, has noted that retrospective studies of intergenerational occupational mobility under-represent fathers whose sons died by time 2 and men in the earlier generation who did not father children.[4] Similar issues will arise in the follow-back study of larger entities, such as formal organizations or even nations. Businesses that went

[2] In a prospective panel study the proportion of time 2 respondents characterized by the rare outcome, although small, will approximate the true proportion of such people in the population. As a consequence, it is legitimate to interpret parameters in a linear model, or in a transformed log-linear model. However, in the follow-back study the relative sizes of the rare-outcome sample and normal population sample are arbitrary, thus making it inappropriate to merge these two samples into a single analysis. In instances of this sort, it is necessary to interpret prediction equations in terms of relative odds and to estimate these equations by means of individual-level logit analysis. The log-linear models developed by Goodman (1975) can be used to estimate these models. For an application of this type of analysis applied to merged samples, see Gortmaker (1979).

[3] By contrast, military records are relatively standardized for all respondents, and are comparatively rich in detail. They cover a substantial percentage of the male population, though with unequal representation of social classes and races. Although sociologists appear to have made little use of these records, they are a data source well worth considering for the longitudinal assessment of a variety of research issues.

[4] The latter omission is not one of censoring, but would lead to a biased sample of the time 1 population. In a *prospective* study, censoring can also arise; for example, children born after time 2 observations are collected would be omitted from the study: their parents would be wrongly classified as not having had children.

bankrupt and nations that were conquered and absorbed by other nations will be omitted from a retrospective study.[5]

The catch-up is a particularly attractive design when the researcher manages to isolate a source of baseline archival data that is especially rich in information. It is, of course, no mean feat to locate data sources in which theoretically interesting information has been collected from a sizable sample of individuals, along with enough identifying information about the respondents (e.g., last known address, or social security number, names of parents, and so forth) to allow a substantial proportion of the target sample to be located.[6] This type of information is typically *not* available, for example, in sets of survey data stored for secondary analysis in data archives. Where data of this sort are available, though, we have wonderful opportunities for follow-up studies. Singer *et al.* (1976), for example, were able to recover enough identifying information in the original Midtown Manhattan Study to conduct a 20-year follow-up with a high rate of sample recovery.

In addition, Robins (1979b) has noted that catch-up designs offer advantages with regard to sampling. There is reason to think that the population at large will underrepresent certain deviant subpopulations of interest to the investigator. As Robins notes:

> troubled people . . . are much less likely to be found in households than is the rest of the population. They have higher death rates, higher rates of incarceration and hospitalization, more time out of the house even when they are officially in residence, and some have no fixed address. Samples chosen from rosters of births and early school records, rosters created before the deviant behavior began, provide a truly representative sample, which when followed, yields more accurate estimates of the number of psychiatrically ill or deviant in a population than do area samples. The more deviant are still less likely to be found at home, but at least the researcher knows that they exist and must be searched for.

At the same time, a new censoring problem arises in the catch-up study. Suppose we start with a baseline population at some earlier time, and wish to compare those members of the population to whom some event (e.g., an arrest, a suicide attempt, an illness) will occur with those to

[5] The sample loss resulting from these processes can be considerable. In the United States, something like one in seven small businesses goes out of business each year. As far as nations are concerned, Charles Tilly (1975b, p. 15) has noted that "The Europe of 1500 included some five hundred more or less independent political units, the Europe of 1900 about twenty-five. Comparing the histories of France, Germany, Spain, Belgium and England . . . for illumination on the processes of state-making weights the whole inquiry toward a certain kind of outcome which was, in fact, quite rare."

[6] Difficulty in locating members of the sample in the follow-up is likely to arise primarily in studying individuals. Organizations or societies tend to be easier to locate. Hunting and gathering bands studied by anthropologists may be the exception that proves the rule.

whom the event will not occur. At time 2, when the researcher "catches
up with" members of the sample selected from the time 1 archives, some
of those to whom the event in question will eventually occur will not yet
have experienced the event in question.[7] Thus they will be misclassified
as individuals to whom the event has not occurred. Generally this
problem will be less important as the interval between time 1 and time 2
increases, but problems of sample attrition may increase as this interval
grows larger.

It may be possible to gain some information about censoring by combining the follow-back and catch-up designs. This can be done by selecting a time 2 sample, following it back to earlier records, and using the
records to select an additional sample that is located and measured at time
2. An analysis of systematic differences between the follow-back sample
and the catch-up sample may then shed some light on the extent to which
censoring has taken place, and enable the researcher to bring recently
developed statistical methods for working with censored data to bear on
the problem (Heckman, 1977; Tuma and Hannan, 1979).

Measurement Intervals

Once having decided on a panel design, the interval of time over which
to space data collection must be decided. This is a decision that must be
made whether new data are to be collected, or data are to be culled from
archives. In the latter case, it may be possible to sidestep this issue by collecting data over many short intervals with little added cost. For example,
a school system will ordinarily have records on the great majority of children enrolled in the school for a period of at least eight years, with entries
spaced annually or semi-annually. But this will not always be true; in
working with census data, for example, one may be limited to data collected decennially. In most cases, researchers will not be able to collect
all the information needed from archives alone, and therefore will have to
engage in a prospective study, or pursue follow-back or catch-up research. All these alternatives will require a decision about time interval
between points of measurement.

As we demonstrated in Chapter 8, a critical consideration in the selection of time intervals between waves concerns the causal lag between the
variables presumably influencing the changes we want to study, and the
criteria. The most central consideration is selecting a causal interval,

[7] Similar problems can arise in studying aggregates. If we carry out a comparison between
nations that have had a revolution with those that have not, some of those we classify in the
latter category would be classified in the former if we were to wait longer.

then, should be the rate of change one expects in the endogenous variables. We cannot study dynamic processes when the amount of time spanning the panel is short relative to the processes of change we want to study. On the other hand, observations collected too far apart may miss short-lived transient effects entirely. Disregarding cost considerations, it is clearly better to have more data than less, a consideration that dictates shorter rather than longer intervals. Most panel studies are multipurpose, and study a variety of causes and changes, which presumably vary among themselves in terms of optimal lag. The optimal choice of a follow-up interval, then—still neglecting cost considerations—is the shortest amount of time that encompasses a *meaningful* amount of change in the criteria of interest to the investigator.[8]

How one defines an amount of change that is *meaningful* will, of course, hinge on substantive considerations unique to the study. The way one determines, before carrying out a study, that a given amount of change will have taken place in a specific interval of time, though, is a more general problem. Often the existing literature can be helpful here. We know, for instance, that screening scales of psychological distress have stabilities of about .50 over a period of two or three years (Kessler and Cleary, 1980), reflecting the fact that these scales measure both variable moods and also comparatively more stable characterological distress. Using the methods of interpolation described in Chapter 9, it is possible to estimate the approximate stability of these scales over shorter intervals of time.

There are many topics, though, about which little is currently known. In situations of this sort, an appeal to the literature will be of little help. When this is true, either of two strategies may be advisable. First, in both catch-up and follow-back studies, it is usually possible to include several age cohorts in the sample. This procedure makes it possible to sample over different time periods by sampling different cohorts. In a study based on school records, for instance, we can obtain grammar school data for respondents who are, at the time of follow-up, 1, 2, 3, or 4 years past high school graduation. By drawing the sample to include members of these different age cohorts, we can have data spanning a variety of time intervals. Although the inability to separate age from cohort effects in comparisons of this kind may prove troublesome for causal analysis, comparisons can nevertheless suggest the order of magnitude of rates of change we can expect.

Second, in a prospective study we have the opportunity to ask respon-

---

[8] If some variables change much more slowly than others, they need not necessarily be measured as frequently as those that change more rapidly.

dents at time 1 some retrospective questions that might help us to determine an appropriate time interval for the follow-up. In most cases we will not have enough faith in responses of this sort to take them at face value. But it frequently happens that we are willing to grant them enough credibility to help us select a time interval for follow-up.

Number of Waves

When at all possible, data should be collected for more than a single follow-up wave (providing three or more waves in total). As we stressed in Chapter 3, these additional waves of data are extremely useful for identifying theoretically interesting models that remain underidentified in two waves. Also, additional waves of data make it possible to determine whether the causal dynamics at work in the system under study are changing over time. We saw in Chapter 7 that the estimation of serially correlated error terms is also facilitated when multiwave data are available. In Chapter 10 it was also shown that single-indicator measurement models can be estimated with three or more waves of data.

Multiwave studies can be particularly important in studying processes about which little a priori information regarding causal lags is available. This is particularly evident in evaluation research, where the effect of an intervention might not be detected until after several follow-up waves.

Even when causal relations are evident in the early waves of observations, further measurements can be useful is assessing long-term tendencies for these effects to decay. It has become common in evaluations of correctional rehabilitation programs, for example, to find that programs show some signs of success (as measured by lower rates of recidivism) over short periods of time, but that these differences erode when longer follow-ups are carried out (Greenberg, 1977). Similarly, the American Headstart program was originally thought to be a success because participants in the program showed IQ gains compared to controls at the end of nursery school. Longer follow-up evaluations of some of these programs found that these gains vanished once the children entered elementary school, reversing the initial impression that the Headstart program permanently eliminated cultural disadvantages in learning.[9]

In prospective studies where researchers must collect data themselves, each additional wave of data typically adds a good deal to the total cost of the research project. In such circumstances, it may be impossible to col-

---

[9] This pessimistic conclusion has recently come under challenge in re-evaluations of Headstart. Our point, though, is methodological, not substantive.

lect data for a number of waves. When this is the case, it may be possible to reinterview the time 1 sample in parts. This approach was utilized by the Center for Epidemiological Studies in its Kansas City and Washington County depression study. Baseline interviews were collected from a probability sample of adults in these two areas, and followed up in stages. Part of the original sample was reinterviewed 3 months after the initial interview; another part was reinterviewed 6 months after the first interview; and the remainder of the sample, after a year had elapsed from the original interview. Each respondent was reinterviewed only once, but the creation of three mutually exclusive subpanels made it possible to study the importance of time-interval.

Parallel subpanels of this sort cannot provide the benefits of model identification that are provided by true multiwave studies. But in allowing analyses to be replicated across varying time intervals, they do provide information about the sensitivity of findings to misspecification of the time lag in the causal model assumed to be operative in the data.

Panel Attrition

A particularly serious problem in some types of prospective panel studies is that of panel loss or attrition.[10] In normal population surveys, it is not uncommon to find response rates of 80%. If nonresponses are independent from one wave to the next, a rate of this magnitude would lead in a three-wave panel study to complete data for all three waves for only 51% of the sample. For four waves, this would drop to 41% of the sample. Typically, panel attrition is not random (Kandel, 1975; Josephson and Rosen, 1978). This means that the panel sample may not only be less complete than a comparable cross-sectional sample, but more biased as well.

Recent changes in government regulations about the protection of human subjects make it even more difficult to complete panels, for it is now much more difficult than it once was to maintain data files linking names, addresses, and other types of identifying information with the answers given by respondents to the time 1 interviews. As a result, many recent studies have used a self-generated code to link records over time,

---

[10] As these difficulties are most acute in survey research, our discussion will focus mainly on the attrition problem as it arises in that context. However, attrition can occur in other types of panel studies as well. For example, the United Nations publishes annual documents reporting a variety of social, economic, and demographic data for member nations, based on information supplied by these nations. It is not uncommon to find that some nations have failed to provide the relevant information in some years, but not others.

an approach that produces completion problems of its own. The respondent to the time 1 interview is asked to generate a code of from five to eight characters (numbers and letters) based on such things as the first two letters of mother's maiden name, followed by the last two numbers of current telephone number, and the third and seventh digits of one's social security number. Most respondents can generate codes of this sort, but they sometimes forget the code later on. Some of the information may change, or mistakes may be made in following the complicated directions required. Thus Rossi and Groves (Groves, 1974) tried to match college students with an eight-digit code based on day and month of respondent's birthday and of respondent's mother's birthday, and found that only 69% of the time 2 respondents could be matched with their time 1 codes.

Two distinct problems are introduced by the fact of sample attrition. The first concerns techniques of following up samples successfully, thereby reducing problems of panel loss. The second concerns methods for taking panel attrition into account in data analysis.

In survey research, the largest contributions to panel loss are likely to be due to failures in locating target respondents in surveys, and interview refusals from respondents who are located. Compared to researchers performing cross-sectional studies, the panel researcher has one advantage: the opportunity to develop a rapport with the respondent over the course of several interviews, thus building a commitment to the research on the part of the respondents.

Keeping this in mind, it is no more than common sense to describe the importance of the project to the respondent during the time 1 interview, to make the interview experience as enjoyable as possible, and generally to set the stage for a successful follow-up. When the interval between data collection points is more than a year, the panel researcher might find it advisable to maintain some interim contact with time 1 respondents. It can be helpful, for example, to send reminder postcards to time 1 respondents sometime before time 2 data are to be collected, both to maintain respondents' involvement with the study and to obtain information on address changes.

Robins (1979b) also suggests that time 1 respondents can be asked to grant the researcher written consent to obtain information from appropriate agencies (such as Internal Revenue Service, police departments, schools, voter registries) that might be useful in locating them.

When the researcher is working from time 1 archival data, or from a time 1 sample collected in the past, these strategies may not be entirely effective, and a more heroic effort may have to be mounted to track down the time 1 respondents. A number of panel researchers have found that such efforts will pay off if the researcher is willing to invest the effort. For example, Clarridge *et al.* (1978) report on their efforts to locate members of a 1957 school sample nearly 20 years later. In the original sample, carried out by Dr. Kenneth Little, more than 30,000 Wisconsin high school seniors filled out a questionnaire. In 1964, Dr. William Sewell used this time 1 baseline to draw a 33% subsample for a follow-up mail survey. In all, 9,007 responses were obtained from a target sample of 10,317. In 1975, an attempt was made to locate as many of the target sample of 10,317 as possible. After making use of a variety of tracing techniques, 97.4% of this sample was located. This includes 99% of the 1964 respondents, and 86.2% of the nonrespondents in 1964. Blacks were located with as high a response rate as whites, and men with as high a rate as women. Out-of-state residents were located with as high a rate as in-state residents. Furthermore, all but the last 1% or 2% of the achieved sample were located in fewer than 90 minutes per respondent.

These results show that difficulties associated with locating follow-up respondents who are part of an earlier time 1 baseline sample need not be a major obstacle to panel completion. This is probably more true in research on the general population than in studies of deviant populations. But nonetheless, it is encouraging.

As we noted above, a much more serious obstacle to panel completion is the problem of *accumulating* panel attrition. Because multiwave studies provide the respondent with more than one opportunity to be missing from the sample, conventional response rates of 80% or so in a cross-section can lead to overall response rates as low as 40% or 50% over three or four waves. Rates as low as this are sufficient to call into question the external validity of findings based on these data.

Several post facto approaches have been used to deal with the problem of nonresponse. The conventional approaches of using pairwise missing matrices for analysis, substituting means or random values for nonresponses in a particular wave, and other sorts of imputation methods not directly tied to the fact that the data being considered are part of a panel have been reviewed in several recent publications (Gleason and Staelin, 1975; Hertel, 1976; Marini *et al.*, 1979), and so will not be reviewed here. Instead, we discuss two methods that seem more uniquely suited to panel data. Both are imputation strategies for creating matrices that are unbiased estimates of complete sample matrices. They do not estimate values on the individual variables for each missing unit of analysis in the panel.

The first technique, due to Rubin (1974, 1977), has been described in nontechnical language by Marini *et al.* (1979). It makes use of the fact that in many panel studies, patterns of nonresponse are nested; that is, a subject who is missing from one wave will be missing in subsequent waves as well.

In reconstructing a covariance matrix for the variables spanning all the waves of data, it would be desirable to make use of the information contained in the earlier waves that is missing in later waves. The Rubin approach allows this to be done by using linear regression to estimate the variances of the variables in later waves from all the data in the earlier waves. This is an unbiased estimate of these variances if the slopes of the variables in later waves on predictors in earlier waves are unbiased. In other words, if we can assume that the causal processes at work in the achieved panel are the same as those at work in the target sample, we can use regression to provide a better estimate of the later variable variances than would be obtained by using the estimate in the achieved samples of these later waves.

If we assume that the covariances estimated in the subsample responding at both times accurately represents the population covariances, the time 1 variances and the imputed time 2 variances can be inserted to yield a complete covariance matrix among the time 1 and time 2 scores. Once this is done, a parallel procedure can be used to construct imputed variances of all time 3 variables, using the imputed time 2 variances for the entire sample in the prediction equations. Means can be estimated in similar fashion by expressing the regression equation in terms of the mean values of all variables.

When the data are multivariate normal, or approximately so, and when the slopes estimated in the restricted over-time samples are unbiased estimates of the slopes that would have been estimated in the complete sample, this procedure yields maximum-likelihood estimates of the variances. Of course, our faith in these estimates depends largely on the extent to which we believe that the slopes are unbiased, and on the number of predictors included in the estimation equations. The longer the duration of the study, the less plausible the first assumption is likely to seem. The Rubin approach, then, seems most appropriate when only two or three waves of data are collected, and the baseline data have been obtained from archival records. In this circumstance, the availability of complete data at time 1 eliminates the nonresponse problem in the estimation of sample means and variances.

In a longer panel, though, or when time 1 data have been collected as part of a sample which itself can have nonresponse, the Rubin method of imputation may not be optimal. Suppose we begin with a target sample of 1000, and that at times 1, 2, 3, and 4 our achieved nested samples are 800, 640, 512, and 410 (each 80% responses of the preceding wave). The Rubin procedure would entail estimating variances for 1000 cases from covariances computed from samples of 640, 512 and 410 for the three contiguous wave pairs. If we have enough faith in the variance estimates based on the 800 time 1 respondents to make use of the Rubin procedure, a more reasonable procedure would be to take the entire sample of 1000 as the target in each wave. With an 80% response rate, we would have an achieved sample of approximately 800 at each time point, and could take variance estimates based on these achieved samples as our estimates of the total sample variances. Since this procedure does not throw away the responses of individuals just because they have once failed to respond in a particular wave, it is reasonable that these separate samples of 800, all drawn from the same target population of 1000, will yield better estimates of the total sample variance than will extrapolations from the time 1 responders based on panels containing between 410 and 640 respondents.

This approach, like Rubin's, rests on the assumption that the same causal relationships prevail among respondents and nonrespondents. Perhaps that is sometimes so, but at times doubt may be in order. We mentioned that some nations fail to report aggregate social, economic, and demographic data regularly for inclusion in United Nations Yearbooks. Often these irregular reporters will be third world nations, whose administrative apparatuses are comparatively undeveloped. Relationships among variables of interest may well be different when the state is comparatively weak, than in nations with more established bureaucracies for collecting and compiling social indicators, and whose states are generally stronger. When nations disappear from a sample because of conquest, or businesses because of bankruptcy or merger, we may wonder whether disappearance from the sample is not the outcome of a causal dynamic that is different from that of survivors.

When suspicions along these lines seem plausible, it is advisable to check for causal homogeneity in the sample before carrying out an imputation scheme. For this purpose, it is by no means adequate to check that the means of individual variables are the same for respondents and nonrespondents—a procedure that many researchers adopt when analyzing cross-sectional data. Instead, the analyst should check that causal relationships are the same for respondents and nonrespondents. When individuals have responded in some waves but not others, a check of this kind can be carried out by utilizing information from those waves where responses have been obtained. Should the samples prove to be inhomogeneous, neither our method nor Rubin's should be adopted for the sample as a whole. Rather, the sample should be disaggregated, and the imputation technique carried out separately for the nonresponding subsample. Parallel but separate causal analyses should then be conducted for the two subsamples.

This procedure is admittedly more complex than the conventional approaches used in cross-sectional analyses. However, the added complexity enables one to learn much more about the social processes that generate one's data than can be learned in cross-sectional analyses. This is so because panels provide a great deal more information—even about units of analysis for which some data are missing—than do cross-sectional designs. As with many of the other methods surveyed in this volume, the richness of panel methods leads to analytical complexity—and through this complexity to a deeper and more complete understanding of the social world.

References

Alwin, D. F., and R. M. Hauser
　1975　"The Decomposition of Effects in Path Analysis." *American Sociological Review*
　　　40:37–47.
Asher, H. B.
　1976　*Causal Modeling.* Beverly Hills: Sage.
Bartholomew, D. J.
　1973　*Stochastic Models for Social Processes.* New York: Wiley.
Bergstrom, A. R. (ed.)
　1976　*Statistical Inference in Continuous Time Economic Models.* New York: Elsevier.
Blumstein, A., J. Cohen, and D. Nagin
　1976　"The Dynamics of a Homeostatic Punishment Process." *Journal of Criminal Law
　　　and Criminology* 68:317–334.
Bohrnstedt, G. W.
　1969　"Observations on the Measurement of Change." Pp. 113–136 in E. F. Borgatta
　　　(ed.), *Sociological Methodology 1969.* San Francisco: Jossey-Bass.
Bohrnstedt, G. W., and G. Marwell
　1978　"The reliability of products of two random variables." Pp. 254–273 in K. F.
　　　Schuessler (ed.) *Sociological Methodology 1978.* San Francisco: Jossey-Bass.
Bollen, K. A.
　1980　"Comparative Measurement of Political Democracy." *American Sociological Re-
　　　view* 45:370–390.
Box, G. E. P., and G. M. Jenkins
　1976　*Time Series Analysis: Forecasting and Control.* Revised edition. San Francisco:
　　　Holden-Day.
Brown, G. W., and T. Harris
　1978　*Social Origins of Depression: A Study of Psychiatric Disorder in Women.* New
　　　York: Free Press.

Campbell, D. T.
  1963 "From Description to Experimentation: Interpreting Trends as Quasi-experiments." Pp. 212–254 in C. W. Harris (ed.), *Problems in the Measurement of Change*. Madison: University of Wisconsin Press.
Campbell, D. T., and J. C. Stanley
  1963 *Experimental and Quasi-Experimental Designs for Research*. Chicago: Rand-McNally.
Campbell, R.
  1978 "Longitudinal Designs in Life Course Research: A Critique." Paper presented to the American Sociological Association.
Chamberlain, G.
  1980 "Analysis of Covariance with Qualitative Data." *Review of Economic Studies* 47:225–238.
Chiang, A. C.
  1974 *Fundamental Methods of Mathematical Economics*. New York: McGraw-Hill.
Clarridge, B. R., L. L. Sheehy, and R. S. Hauser
  1978 "Tracing Members of a Panel: A 17-Year Follow Up." Pp. 185–203 in K. F. Schuessler (ed.), *Sociological Methodology 1978*. San Francisco: Jossey-Bass.
Coleman, J. S.
  1964 *Models of Change and Response Uncertainty*. Englewood Cliffs, N.J.: Prentice-Hall.
  1968 "The Mathematical Study of Change." Pp. 428–478 in H. M. Blalock, Jr. and A. B. Blalock (eds.), *Methodology in Social Research*. New York: McGraw Hill.
Collver, A., and M. Semyonov
  1979 "Suburban Change and Persistence." *American Sociological Review* 44:480–486.
Cook, T. D., and D. T. Campbell
  1976 "The Design and Conduct of Quasi-experiments and True Experiments in Field Settings." Pp. 223–326 in M. D. Dunnette (ed.), *Handbook of Industrial and Organizational Psychology*. Chicago: Rand McNally.
  1979 Quasi-Experimentation: Design and Analysis: Issues for Field Settings. Chicago: Rand McNally.
Costner, H. L.
  1969 "Theory, Deduction and Rules of Correspondence." *American Journal of Sociology* 75:245–263.
Costner, H. L., and R. Schoenberg
  1973 "Diagnosing Indicator Ills in Multiple Indicator Models." Pp. 167–199 in A. S. Goldberger and O. D. Duncan (eds.), *Structural Equation Models in the Social Sciences*. New York: Seminar Press.
Crano, W. D., D. A. Kenny, and D. T. Campbell
  1972 "Does Intelligence Cause Achievement? A Cross-Lagged Panel Analysis." *Journal of Educational Psychology* 63:258–275.
Cronbach, L. J., and L. Furby
  1969 "How to Measure Change—or Should We?" *Psychological Bulletin* 74:68–80. Also errata, n.p.
Dohrenwend, B. P., and B. S. Dohrenwend
  1969 *Social Status and Psychological Disorder*. New York: Wiley.
Duncan, O. D.
  1966a "Path Analysis: Sociological Examples." *American Journal of Sociology* 72:1–16.
  1966b "Methodological Issues in the Analysis of Social Mobility." Pp. 51–97 in N.

Smelser and S. M. Lipset (eds.), *Social Structure and Social Mobility in Economic Development.* Chicago: Aldine.

1969a   "Some Linear Models for Two-Wave, Two-Variable Panel Analysis." *Psychological Bulletin* 72:177–182.

1969b   "Contingencies in Constructing Causal Models." Pp. 74–112 in Edgar F. Borgatta (ed.), *Sociological Methodology 1969.* San Francisco: Jossey-Bass.

1972   "Unmeasured Variables in Linear Models for Panel Analysis." Pp. 36–82 in H. L. Costner (ed.), *Sociological Methodology 1972.* San Francisco: Jossey-Bass.

1975   *Introduction to Structural Equation Models.* New York: Academic Press.

1980   "Testing Key Hypotheses in Panel Analysis." Pp. 279–289 in K. F. Schuessler (ed.), *Sociological Methodology 1980.* San Francisco: Jossey-Bass.

Eaton, W. W.
1978   "Life Events, Social Supports, and Psychiatric Symptons: A Re-Analysis of the New Haven Data." *Journal of Health and Social Behavior* 19:230–234.

Faith, R. E.
1973   "Regression in Panel Data." Unpublished paper. Ann Arbor: University of Michigan Survey Research Center.

Fararo, T. J.
1968   "Stochastic Processes." Pp. 245–262 in E. F. Borgatta (ed.), *Sociological Methodology 1969.* San Francisco: Jossey-Bass.

1973   *Mathematical Sociology: An Introduction to Fundamentals.* New York: Wiley.

Fisher, F. M.
1966   *The Identification Problem in Econometrics.* New York: McGraw-Hill.

Fisher, F., and D. Nagin
1978   "On the Feasibility of Identifying the Crime Function in a Simultaneous Model of Crime Rates and Sanction Levels." Pp. 250–312 in A. Blumstein, J. Cohen, and D. Nagin (eds.), *Deterrence and Incapacitation: Estimating the Effects of Criminal Sanctions on Crime Rates.* Washington, D.C.: National Academy of Sciences.

Gleason, T. C., and R. Staelin
1975   "A Proposal for Handling Missing Data." *Psychometrika* 40:229–252.

Glock, C. Y.
1955   "Some Applications of the Panel Method to the Study of Change." Pp. 242–250 in P. F. Lazarsfeld and M. Rosenberg (eds.), *The Language of Social Research.* New York: Free Press.

Goldberg, S.
1958   *Introduction to Difference Equations.* New York: Wiley.

Goldberger, A. S.
1971   "Econometrics and Psychometrics: A Survey of Communalities." *Psychometrika* 36:83–107.

Goldberger, A. S., and O. D. Duncan
1973   *Structural Equation Models in the Social Sciences.* New York: Academic Press.

Goodman, L. A.
1975   "The Relationship between the Modified and More Usual Regression Approaches to the Analysis of Dichotomous Variables." Pp. 83–110 in D. R. Heise (ed.), *Sociological Methodology 1976.* San Francisco: Jossey-Bass.

Gordon, R. A.
1968   "Issues in Multiple Regression." *American Journal of Sociology* 73:592–616.

Gortmaker, S. L.
1979   "Poverty and Infant Mortality in the United States." *American Sociological Review* 44:28–297.

Greenberg, D. F.
  1979 *Mathematical Criminology*. New Brunswick, N.J.: Rutgers University Press.
Greenberg, D. F., R. C. Kessler, and C. H. Logan
  1979 "A Panel Model of Crime Rates and Arrest Rates." *Americal Sociological Review* 44:843–850.
Groves, W. E.
  1974 "Patterns of College Drug Use and Life Styles." Pp. 241–275 in E. Josephson and E. E. Carroll (eds.) *Drug Use: Epidemiological and Sociological Approaches.* Washington, D.C.: Hemisphere.
Halaby, C. N., and M. E. Sobel
  1979 "Mobility Effects in the Workplace." *American Journal of Sociology* 85:385–416.
Hannan, M. T., and A. A. Young
  1977 "Estimation in Panel Models: Results on Pooling Cross-Sections and Time Series. Pp. 52–83 in D. R. Heise (ed.), *Sociological Methodology 1977*. San Francisco: Jossey-Bass.
Hannan, M. T., and N. B. Tuma
  1979 "Methods for Temporal Analysis." Pp. 303–328 in A. Inkeles, J. Coleman, and R. H. Turner (eds.), *Annual Review of Sociology*. Palo Alto: Annual Reviews.
  forthcoming *Social Dynamics: Models and Methods*. New York: Academic Press.
Hanushek, E. A., and J. E. Jackson
  1977 *Statistical Methods for Social Scientists*. New York: Academic Press.
Hargens, L. L., B. F. Raskin, and P. D. Allison
  1976 "Problems in Estimating Measurement Error from Panel Data: An Example Involving the Measurement of Scientific Productivity." *Sociological Methods and Research* 4:439–58.
Heckman, J. J.
  1977 "Sample Selection Bias as a Specification Error." Rand Corporation Report R-1984.
Heinlein, R.
  1966 *The Moon is a Harsh Mistress*. New York: Putnam.
Heise, D. R.
  1969 "Separating Reliability and Stability in Test-Retest Correlation." *American Sociological Review* 34:93–101.
  1970 "Causal Inference from Panel Data." Pp. 3–27 in E. F. Borgatta and G. W. Bohrnstedt (eds.), *Sociological Methodology 1970*. San Francisco: Jossey-Bass.
Hertel, B. R.
  1976 "Minimizing Error Variance Introduced by Missing Data Routines in Survey Analysis." *Sociological Methods and Research* 4:459–474.
Hibbs, D. A., Jr.
  1974 "Problems of Statistical Estimation and Causal Inference in Time-Series Regression Models." Pp. 252–308 in H. L. Costner (ed.), *Sociological Methodology 1973–1974*. San Francisco: Jossey-Bass.
Hope, K.
  1975 "Models of Status Inconsistency and Social Mobility Effects." *American Sociological Review* 40:336–343.
Humphreys, L. G., and J. Stubbs
  1977 "A Longitudinal Analysis of Teacher Expectation, Student Expectation, and Student Achievement." *Journal of Educational Measurement* 14:261–270.
Jackman, R. W.
  1980 "A Note on the Measurement of Growth Rates in Cross-National Research." *American Journal of Sociology* 86:604–617.

Jöreskog, K. G.
1973   "A General Method for Estimating a Linear Structural Equation System." Pp. 85–112 in A. S. Goldberger and O. D. Duncan (eds.), *Structural Equation Models in the Social Sciences*. New York: Seminar Press.

Jöreskog, K. G., and D. Sörbom
1975   "LISREL III: Estimation and Linear Structural Equation Systems by Maximum-Likelihood Methods." Department of Statistics, University of Uppsala.
1976   "Statistical models and methods for Test-Retest Situations." Pp. 135–157 in D. N. M. Gruijter and L. J. van der Kamp (eds.), *Advances in Psychological and Educational Measurement*. New York: Wiley.
1977   Statistical Models and Methods for Analysis of Longitudinal Data." Pp. 285–325 in D. J. Aigner and A. S. Goldberger (eds.), *Latent Variables in Socioeconomic Models*. Amsterdam: North-Holland.
1979   "LISREL IV: Analysis of Linear Structural Relationships by the Method of Maximum Likelihood." Department of Statistics, University of Uppsala, Sweden.

Johnston, J.
1972   *Econometric Methods*. New York: McGraw-Hill.

Josephson, E., and M. Rosen
1978   "Panel Loss in a High School Drug Study." Pp. 115–136 in D. B. Kandel (ed.), *Longitudinal Research on Drug Use*. Washington, D.C.: Hemisphere.

Kahle, L. R., and J. J. Berman
1979   "Attitudes Cause Behaviors: A Cross-Lagged Panel Analysis." *Journal of Personality and Social Psychology* 37:315–321.

Kaplan, H. B.
1975   "Increase in Self-Rejection as an Antecedent of Deviant Responses." *Journal of Youth and Adolescence* 4:281–292.

Kaufman, R. L.
1976   "The Solution and Interpretation of Differential Equation Models (Comment on Freeman and Hannan)." *American Sociological Review* 41:746–748.

Kenny, D. A.
1973   "Cross-lagged and Synchronous Common Factors in Panel Data." Pp. 153–165 in A. S. Goldberger and O. D. Duncan (eds.), *Structural Equation Models in the Social Sciences*. New York: Seminar Press.
1975   "Cross-Lagged Panel Correlation: A Test for Spuriousness." *Psychological Bulletin* 82:887–903.
1979   *Correlation and Causality*. New York: Wiley-Interscience.

Kenny, D. A., and J. M. Harackiewicz
1979   "Cross-lagged Panel Correlation: Practice and Promise." *Journal of Applied Psychology* 64:372–379.

Kessler, R. C.
1977a   "Rethinking the 16-Fold Table Problem." *Social Science Research* 6:84–107.
1977b   "The Use of Change Scores as Criteria in Longitudinal Survey Research." *Quality and Quantity* 11:43–66.

Kessler, R. C., and P. D. Cleary
1980   "Social Class and Psychological Distress." *American Sociological Review* 45:463–78.

Kidder, L. H., R. L. Kidder, and P. Synderman
1974   "Do Police Cause Crime? A Cross-Lagged Panel Analysis." Unpublished Paper. Temple University.

Kohn, M. L., and Carmi Schooler
1978   "The Reciprocal Effects of the Substantive Complexity of Work and Intellectual

Flexibility: A Longitudinal Assessment." *American Journal of Sociology* 84:24–52.

Land, K. C., and M. Felson
1978 "Sensitivity Analysis of Arbitrarily Identified Simultaneous-Equation Models." *Sociological Methods and Research* 6:283–307.

Lazarsfeld, P. F.
1948 "The Use of Panels in Social Research." *Proceedings of the American Philosophical Society,* and reprinted in Pp. 330–337 of P. F. Lazarsfeld, A. K. Pasanella, and M. Rosenberg, *Continuities in the Language of Social Research.* New York: Free Press.

Lazarsfeld, P. F., B. Berelson, and H. Gaudet
1948 *The People's Choice.* New York: Columbia University Press.

Lefkowitz, M., L. D. Eron, L. O. Walker, and L. Rowell Huesman
1970 *Growing Up to Be Violent: A Longitudinal Study of the Development of Aggression.* New York: Penguin.

Liem, R., and J. Liem
1978 Social Class and Mental Illness Reconsidered: The Role of Economic Stress and Social Support." *Journal of Health and Social Behavior* 19:139–156.

Logan, C. H.
1975 "Arrest Rates and Deterrence." *Social Science Quarterly* 56:376–389.

Long, J. S.
1976 "Estimation and Hypothesis Testing in Linear Models Containing Measurement Error: A Review of Jöreskog's Model for the Analysis of Covariance Structures." *Sociological Methods and Research* 5:157–206.

Long, S. B.
1980 "The Continuing Debate over the Use of Ratio Varaibles: Fact and Fiction." Pp. 37–67 in K. F. Schuessler (ed.), *Sociological Methodology 1980.* San Francisco: Jossey-Bass.

McClendon, McK. J.
1977 "Structural and Exchange Components of Vertical Mobility." *American Sociological Review* 42:56–74.

Marini, M. M., A. R. Olsen, and D. B. Rubin
1979 "Maximum-Likelihood Estimation in Panel Studies with Missing Data." Pp. 314–357 in K. Schuessler (ed.), *Sociological Methodology 1980.* San Francisco: Jossey-Bass.

Markus, G.
1980 *Models for the Analysis of Panel Data.* Beverly Hills: Sage.

Miller, A. D.
1971 "Logic of Causal Analysis: From Experimental to Nonexperimental Designs." Pp. 273–294 in H. M. Blalock, Jr. (ed.), *Causal Models in the Social Sciences.* Chicago: Aldine.

Mortimer, J. T., and J. Lorence
1979 "Work Experience and Occupational Value Socialization: A Longitudinal Study," *American Journal of Sociology* 84:1361–1385.

Myers, J. K., J. J. Lindenthal, and M. P. Peppar
1974 "Social Class, Life Events and Psychiatric Symptons: A Longitudinal Study." Pp. 191–206 in B. P. Dohrenwend and B. S. Dohrenwend (eds.), *Stressful Life Events.* New York: Wiley.

Nagin, D.
1978 "General Deterrence: A Review of the Empirical Evidence." Pp. 110–174 in A. Blumstein, J. Cohen, and D. Nagin (eds.), *Deterrence and Incapacitation: Esti-*

mating the Effects of Criminal Sanctions on Crime Rates. Washington, D.C.: National Academy of Sciences.

Namboodiri, H. K., L. F. Carter, and H. M. Blalock, Jr.
1975   *Applied Multivariate Analysis and Experimental Design.* New York: McGraw-Hill.

Nielson, F., and R. Rosenfeld
1978   "Substantive Interpretations of Differential Equations Models." Unpublished paper.

Papanek, G.
1973   "Aid, Foreign Private Investment, Savings and Growth in Lesser Developed Countries." *Journal of Political Economy* 81:120–130.

Pelz, D. C., and F. M. Andrews
1964   "Detecting Causal Priorities in Panel Study Data." *American Sociological Review* 29:836–848.

Pelz, D. C., and R. E. Faith
1971   *Causal Connections in Educational Panel Data.* Final Report. Washington, D.C.: Department of Health, Education and Welfare, Office of Education.

Pelz, D. C., and R. E. Faith
1973   "Detecting Causal Connections in Panel Data." Unpublished paper, University of Michigan Survey Research Center.

Pelz, D. C., and R. A. Lew
1970   "Heise's Causal Model Applied." Pp. 28–37 in E. F. Borgatta (ed.), *Sociological Methodology 1970.* San Francisco: Jossey-Bass.

Pelz, D., S. Magliveras, and R. A. Lew
1968   "Correlational Properties of Simulated Panel Data with Causal Connections between Two Variables." Unpublished paper, University of Michigan Survey Research Center.

Pontell, H. N.
1978   "Deterrence: Theory versus Practice." *Criminology* 16:3–22.

Rao, P., and R. L. Miller
1971   *Applied Econometrics.* Belmont, Calif.: Wadsworth.

Robins, L. N.
1979a   "Follow-up Studies of Behavior Disorders in Children." Pp. 414–450 in H. C. Quay and J. S. Werry (eds.), *Psychopathological Disorders of Children,* Second edition. New York: Wiley.
1979b   "Longitudinal Methods in the Study of Normal and Pathological Development." Pp. 627–684 in K. P. Kisker, J. E. Meyer, C. Müller, and E. Stromgren (eds.), *Psychiatrie der Gerenwart,* Vol. 1. 2nd ed. Heidelberg: Spring-Verlag.

Rogosa, D.
1979   "Causal Models in Longitudinal Research: Rationale, Formulation, and Interpretation." Pp. 263–302 in J. R. Nesselroode and P. B. Baltes (eds.), *Longitudinal Research in Human Development: Design and Analysis.* New York: Academic Press.
1980   "A Critique of Cross-Lagged Correlation." *Psychological Bulletin* 88:245–58.

Rozelle, R. M., and D. T. Campbell
1969   "More Plausible Rival Hypotheses in the Cross-Lagged Panel Correlation Technique." *Psychological Bulletin* 71:74–80.

Rubin, D. B.
1974   "Characterizing the Estimation of Parameters in Incomplete Data Problems." *Journal of the American Statistical Association* 69:467–474.
1977   "Formalizing Subjective Notions about the Effect of Nonrespondents in Sample Surveys." *Journal of the American Statistical Association* 72:538–543.

Sims, H. P., and D. A. Wilkerson

1977 "Time-Lags in Cross-Lag Correlation Studies: A Computer Simulation." *Decision Sciences* 8:630–644.

Singer, B., and S. Spilerman
1976 "Some Methodological Issues in the Analysis of Longitudinal Surveys." *Annals of Economic and Social Measurement* 5:447–474.
1979 "Mathematical Representations of Development Theories." Pp. 157–177 in J. Nesslroade and P. B. Baltes (eds.), *Longitudinal Research in the Study of Behavior and Development.* New York: Academic Press.

Singer, E., S. M. Cohen, R. Garfinkel, and L. Srole
1976 "Replicating Psychiatric Ratings through Multiple Regression Analysis: The Midtown Manhattan Restudy." *Journal of Health and Social Behavior* 17:376–387.

Sontag, L. W.
1971 "The History of Longitudinal Research: Implications for the Future." *Child Development* 42:987–1002.

Sörbom, D.
1975 "Detection of Correlated Errors in Longitudinal Data," *British Journal of Mathematical and Statistical Psychology* 28:138–151.
1976 "A Statistical Model for the Measurement of Change in True Scores." Pp. 159–169 in D. N. M. Gruijter and L. J. van der Kamp (eds.), *Advances in Psychological and Educational Measurement.* New York: Wiley.

Sørenson, A.
1973 "Causal Analysis of Cross-Sectional and Over-Time Data: With Special Reference to the Study of the Occupational Achievement Process." Madison: University of Wisconsin Institute for Research on Poverty Discussion Paper.

Southwood, K. E.
1978 "Substantive Theory and Statistical Interaction: Five Models." *American Sociological Review* 83:1154–1203.

Spilerman, S.
1971 "The Causes of Racial Disturbances: Tests of an Explanation." *American Sociological Review* 36:427–442.

Stevenson, P.
1972 "External Economic Variables Influencing the Economic Growth Rate of Seven Major Latin American Nations." *Canadian Review of Sociology and Anthropology* 9:347–356.

Stimson, J., E. G. Carmines, and R. A. Zeller
1978 "Interpreting Polynomial Regression," *Sociological Methods and Research* 6:515–523.

Taylor, H. F., and C. A. Hornung
1979 "On a General Model for Social and Cognitive Consistency," *Sociological Methods and Research* 7:259–287.

Tilly, C.
1975 *The Formation of National States in Western Europe.* Princeton, N.J.: Princeton University Press.

Tilly, C., L. A. Tilly, and R. Tilly
1975 *The Rebellious Century.* Cambridge: Harvard University Press.

Tittle, C. R., and A. R. Rowe
1974 "Certainty of Arrest and Crime Rates: A Further Test of the Deterrence Hypothesis." *Social Forces* 52:455–462.

Tuma, N. B., and M. T. Hannan
1979 "Approaches to the Censoring Problem in Analysis of Event Histories." Pp.

209–240 in K. F. Schuessler (ed.), *Sociological Methodology 1979*. San Francisco: Jossey-Bass.

Turner, R. J.
1968   "Social Mobility and Schizophrenia." *Journal on Health and Social Behavior* 9:194–203.

Turner, R. J., and M. O. Wagenfeld
1967   "Occupational Mobility and Schizophrenia: An Assessment of the Social Causation and Social Selection Hypotheses." *American Sociological Review* 32:104–113.

Werts, C. E., and R. L. Linn
1975   Study of Academic Growth Using Simplex Models. Final Report, U.S. Department of Health, Education and Welfare. Washington, D.C.: National Institute of Education.

Wheaton, B.
1978   "The Sociogenesis of Psychological Disorder: Reexamining the Causal Issues with Longitudinal Data." *American Sociological Review* 43:383–403.

Wheaton, B., B. Muthén, D. F. Alwin, and G. F. Summers
1977   "Assessing Reliability and Stability in Panel Models." Pp. 84–136 in D. R. Heise (ed.), *Sociological Methodology 1977*. San Francisco: Jossey-Bass.

Wiggins, L. M.
1973   *Panel Analysis: Latent Probability Models for Attitude and Behavior Processes.* San Francisco: Jossey-Bass.

Wiley, D. E., and J. A. Wiley
1970   "The Estimation of Measurement Error in Panel Data." *American Sociological Review* 35:112–117.

Wiley, J., and M. G. Wiley
1974   "A Note on Correlated Errors in Repeated Measurements." *Sociological Methods and Research* 3:172–188.

Zurcher, L. A., and K. L. Wilson
1979   "Status Inconsistency and the Hope Technique, II: A Linear Hypothesis about Status Enhancement, Status Detraction and Satisfaction with Membership." *Social Forces* 57:1248–1264.

Subject Index

QUANTITATIVE STUDIES IN SOCIAL RELATIONS

Consulting Editor: Peter H. Rossi

UNIVERSITY OF MASSACHUSETTS
AMHERST, MASSACHUSETTS

Richard F. Curtis and Elton F. Jackson, INEQUALITY IN AMERICAN COMMUNITIES

Richard A. Berk, Harold Brackman, and Selma Lesser, A MEASURE OF JUSTICE: An Empirical Study of Changes in the California Penal Code, 1955–1971

Samuel Leinhardt (Ed.), SOCIAL NETWORKS: A Developing Paradigm

Donald J. Treiman, OCCUPATIONAL PRESTIGE IN COMPARATIVE PERSPECTIVE

Beverly Duncan and Otis Dudley Duncan, SEX TYPING AND SOCIAL ROLES: A Research Report

N. Krishnan Namboodiri (Ed.), SURVEY SAMPLING AND MEASURE-MENT

Robert M. Groves and Robert L. Kahn, SURVEYS BY TELEPHONE: A National Comparison with Personal Interviews

Peter H. Rossi, Richard A. Berk, and Kenneth J. Lenihan, MONEY, WORK, AND CRIME: Experimental Evidence

Walter Williams, GOVERNMENT BY AGENCY: Lessons from the Social Program Grants-in-Aid Experience

Juan E. Mezzich and Herbert Solomon, TAXONOMY AND BEHAVIORAL SCIENCE

Zev Klein and Yohanan Eshel, INTEGRATING JERUSALEM SCHOOLS

Philip K. Robins, Robert G. Spiegelman, Samuel Weiner, and Joseph G. Bell (Eds.), A GUARANTEED ANNUAL INCOME: Evidence from a Social Experiment

James Alan Fox (Ed.), MODELS IN QUANTITATIVE CRIMINOLOGY

James Alan Fox (Ed.), METHODS IN QUANTITATIVE CRIMINOLOGY

Ivar Berg (Ed.), SOCIOLOGICAL PERSPECTIVES ON LABOR MAR-KETS

Ronald C. Kessler and David F. Greenberg, LINEAR PANEL ANALYSIS: Models of Quantitative Change

In Preparation

Bruce Jacobs, THE POLITICAL ECONOMY OF ORGANIZATIONAL CHANGE: Urban Institutional Response to the War on Poverty

Michael E. Sobel, LIFESTYLE AND SOCIAL STRUCTURE: Concepts, Definitions, Analyses

Howard Schuman and Stanley Presser, QUESTIONS AND ANSWERS IN ATTITUDE SURVEYS: Experiments on Question Form, Wording, and Context